Adam B. Levy
**Optimal Control**

# Also of Interest

*Computational Physics*
*With Worked Out Examples in FORTRAN® and MATLAB®*
Michael Bestehorn, 2023
ISBN 978-3-11-078236-3, e-ISBN (PDF) 978-3-11-078252-3

*Optimization and Control for Partial Differential Equations*
*Uncertainty quantification, open and closed-loop control, and shape*
*optimization*
Edited by Roland Herzog, Matthias Heinkenschloss, Dante Kalise,
Georg Stadler, Emmanuel Trélat, 2022
ISBN 978-3-11-069596-0, e-ISBN (PDF) 978-3-11-069598-4
in: Radon Series on Computational and Applied Mathematics
ISSN 1865-3707

*Geodesy*
Wolfgang Torge, Jürgen Müller, Roland Pail, 2023
ISBN 978-3-11-072329-8, e-ISBN (PDF) 978-3-11-072330-4

*Topics in Complex Analysis*
Joel L. Schiff, 2022
ISBN 978-3-11-075769-9, e-ISBN (PDF) 978-3-11-075782-8
in: De Gruyter Studies in Mathematics
ISSN 0179-0986

*Partial Differential Equations*
*An Unhurried Introduction*
Vladimir A. Tolstykh, 2020
ISBN 978-3-11-067724-9, e-ISBN (PDF) 978-3-11-067725-6

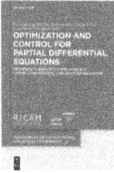

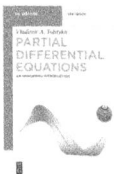

Adam B. Levy

# Optimal Control

From Variations to Nanosatellites

DE GRUYTER

**Mathematics Subject Classification 2020**
Primary: 49N05; Secondary: 49N99

**Author**
Dr. Adam B. Levy
Bowdoin College
Department of Mathematics
8600 College Station
Brunswick ME 04011
USA
alevy@bowdoin.edu

ISBN 978-3-11-128983-0
e-ISBN (PDF) 978-3-11-129015-7
e-ISBN (EPUB) 978-3-11-129050-8

**Library of Congress Control Number: 2023937376**

**Bibliographic information published by the Deutsche Nationalbibliothek**
The Deutsche Nationalbibliothek lists this publication in the Deutsche Nationalbibliografie;
detailed bibliographic data are available on the Internet at http://dnb.dnb.de.

Cover image: Rick_Jo / iStock / Getty Images Plus
Typesetting: VTeX UAB, Lithuania
Printing and binding: CPI books GmbH, Leck

www.degruyter.com

To my wife Sarah and kids, Jonah and Emma. And to my Dad

# Preface

Optimal control is an important applied mathematical subject where an agent has some control over underlying dynamics and seeks the strategy that optimizes some significant quantity. This subject synthesizes ideas from optimization and differential equations, so it is an ideal applied mathematics topic for advanced undergraduates and early graduate students looking to grow and enhance their understanding of these more introductory applied topics. This text provides an accessible introduction to optimal control, covering its historical roots in the calculus of variations and developing the modern ideas as a natural extension of those roots. This approach simplifies the theory and invites a wider audience to access this important material.

To facilitate understanding, learning-by-doing activities are embedded throughout the text where they are relevant. One form of these activities are Problems for the reader to solve as they go (or for groups of students to solve together). In addition, there are Questions for the reader to ponder and answer relatively quickly, and there are Lab Activities, which invite the use of application software to aid investigation. Any application software that solves differential equations will support these activities, and coding suggestions are provided here for using Python or Mathematica to complete the Lab Activities. Python is one of the most popular open-source programming languages in the world and is therefore widely available. Mathematica is a popular commercial symbolic mathematical computation program. There are also chapter-ending Exercises, on which the readers can test their understanding. Answers to the Questions and complete solutions to the Problems and Lab Activities can be found in Appendices.

There are broad applications for optimal control across engineering, operations research, and the natural and social sciences. These include problems in medicine, agriculture, counterterrorism, traffic flow, electric vehicles, aerospace, food security, athletics, neuroscience, conservation, advertising, bioengineering, economics, and robotics. The finale to this text is an invitation to read current research in an application area of interest, and the balance of the text leading up to that finale prepares the reader to gain a solid understanding of the current research they read.

## Organization

The first third of this text is an introduction to the calculus of variations, which is the classical precursor to optimal control. The calculus of variations was invented at the same time as what we now call "calculus", and many of the same people contributed to these parallel developments. We motivate our investigation in this part of the text with two classical problems in the calculus of variations, and we learn the fundamental "Euler–Lagrange equation", which identifies optimizer candidates. We also learn a "necessary" condition, which can rule out optimizer candidates, and a "sufficient" condition, which can help us narrow the candidate pool to an actual optimizer. Along the

https://doi.org/10.1515/9783111290157-201

way, we discuss how the function class from which the candidates are drawn can affect whether optimizers even exist.

The rest of the text covers optimal control, which we introduce as an extension of the calculus of variations to accommodate an agent with some control over underlying dynamics. We discuss "discrete-time" optimal control problems (where time is measured in periods) and solve one using the famous "dynamic programming" method [1]. However, most of the second part of the text focusses on continuous-time optimal control problems (where time is measured on a continuum), including problems with multiple "states" that obey linear dynamics. Such multiple-state control problems are characterized by the type of eigenvalues associated with their dynamics since those eigenvalues dictate the behavior of the states. We will see how the Euler–Lagrange equation from the calculus of variations can be traced to an important equation in optimal control, and we will give an intuitive justification for "Pontryagin's principle" identifying optimizing candidates in optimal control problems. Different versions of a nanosatellite control problem will be used throughout this part of the text to motivate, illustrate, and provide intuition for the theory. This nanosatellite problem thus provides the companion bookend to the calculus of variations, and these two bookends together motivate the subtitle "from variations to nanosatellites" of this text.

## Prerequisites

The most significant prerequisite for understanding this material is previous experience with solving differential equations and systems of differential equations. Solving a calculus of variations problem typically requires finding a solution to a differential equation, and our study of optimal control is organized by different types of eigenvalues that dictate the dynamics underlying the problems.

## Outcomes

By working through this text, the reader should
–   Appreciate the significance of integral optimization problems.
–   Understand the origin of the classical Euler–Lagrange equation in the calculus of variations.
–   See how calculus of variations problems sit within a broader framework of optimal control.
–   Connect the solution technique in optimal control to the Euler–Lagrange equation.
–   Have an intuitive understanding of Pontryagin's principle.
–   Learn to apply the optimal control solution technique to solve a variety of problems.
–   Understand an interesting application of optimal control theory to a current problem in engineering, operations research, or the natural and social sciences.

# Acknowledgments

This text was developed over several decades of teaching this material to students at Bowdoin College, who have made many important contributions during this development. Existing texts on this material were also very influential on the content, organization, and level of the present text. Of particular note in this regard is [2], followed by the more recent [3]. Some of the examples, problems, and exercises herein are extracted or adapted from these and other sources.

This text appears in its final form as a result of very professional and efficient shepparding from Ranis Ibragimov, Ute Skambraks, and their team at De Gruyter, as well as from Vilma Vaičeliūnienė at VTeX.

https://doi.org/10.1515/9783111290157-202

# Contents

## Part I:  Calculus of variations

## Part II:  Optimal control

# List of Tables

https://doi.org/10.1515/9783111290157-203

# List of Figures

https://doi.org/10.1515/9783111290157-204

Part I: **Calculus of variations**

# 1 The brachistochrone

When modern calculus was being developed during the seventeenth century, the Swiss mathematician Johan Bernoulli publicly posed a challenge problem for his peers to solve [4, Chapter 7]:

> Given two points A and B in a vertical plane, what is the curve traced out by a point acted on only by gravity, which starts at A and reaches B in the shortest time.

This situation is illustrated in Figure 1.1. The story goes that Bernoulli already knew the solution to this problem and taunted his rivals (with thinly veiled references, in particular, to Issac Newton) that they would not be able to solve it. The counterstory from Newton's camp is that he solved it easily after he got home from his day-job of managing "The Great Recoinage of 1696" (replacing one set of coins in Britain with another) as Warden of the Royal Mint.

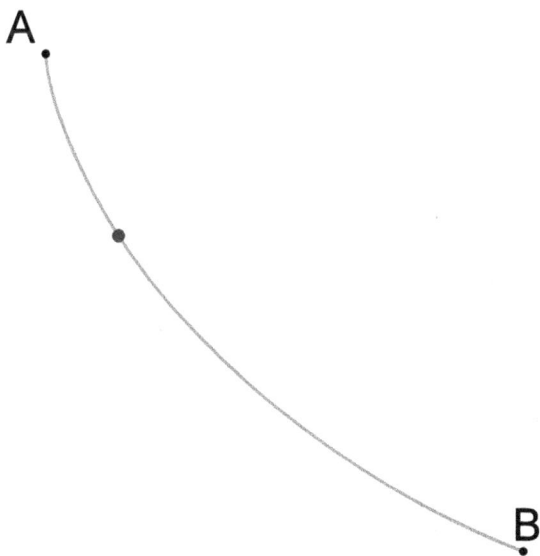

**Figure 1.1:** Brachistochrone.

This problem is called the "brachistochrone" problem from the Greek words "brachistos" (the shortest) and "chronos" (time), and it was solved by other mathematicians in addition to Isaac Newton and Johann Bernoulli. The brachistochrone problem soon led to other problems of its type, and eventually—via Leonhard Euler and Joseph-Louis de Lagrange—to a new field of mathematics called "Calculus of Variations".

https://doi.org/10.1515/9783111290157-001

## 1.1 Calculus of variations formulation

The brachistochrone problem can be restated as finding the function $x(t)$ whose graph not only goes through the two points $A = (0,1)$ and $B = (1,0)$ as in Figure 1.2, but also gives the path of least time.

graph of x(t)

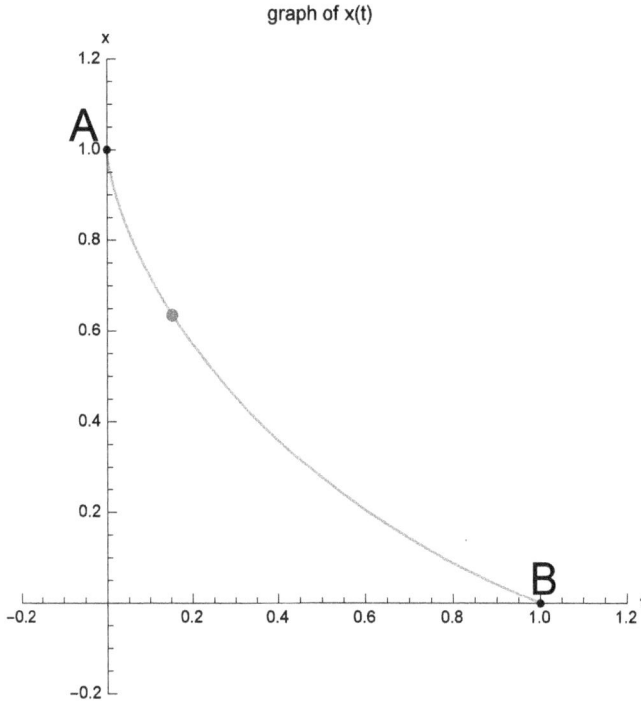

**Figure 1.2:** Brachistochrone on standard axes.

The time it takes for a point to move via gravity along the graph of any such $x(t)$ can be obtained by integrating the inverse of the speed with respect to arclength, which in this case simplifies to a constant multiple of the integral "functional" $J[x]$ defined by

$$J[x] = \int_0^1 \sqrt{\frac{1 + \dot{x}^2}{1 - x}}\, dt \tag{1.1}$$

in terms of the derivative $\dot{x} = \dot{x}(t)$ with respect to $t$ of the function $x = x(t)$.

### 1.1.1 Functionals

Recall that functions take input variables (points) and generate output values (numbers). The "al" at the end of the word *functional* indicates that the input variables are functions $x(t)$ themselves, and we use the square brackets "$[x]$" to signal this distinction. Thus the integral $J[x]$ defined in (1.1) is a function-like object that generates an output value (i. e., the value of the integral) from any function $x(t)$ to which it is applied.

**Problem 1.1.**
(a) *Simplify the output values $J[x_\epsilon]$ for the integral functional (1.1) applied to functions of the form*

$$x_\epsilon(t) = 1 - t^\epsilon \quad \text{for } \epsilon > 0$$

*as much as possible (without computing the integral).*
(b) *Compute the output value $J[x_1]$ and describe the graph of $x_1(t)$.*

### 1.1.2 Admissibility

Since the integral functional $J[x]$ defined in (1.1) measures (a constant multiple of) the time it takes for a point to move via gravity along the graph of $x(t)$, it follows that the brachistochrone problem amounts to minimizing $J[x]$ over functions $x(t)$ satisfying $x(0) = 1$ and $x(1) = 0$ (ensuring that the graph of $x(t)$ connects the points $A = (0,1)$ and $B = (1,0)$). The term *admissible* is used to identify the functions that we consider for the minimization, so the admissible functions $x(t)$ for the brachistochrone problem must at least satisfy $x(0) = 1$ and $x(1) = 0$.

The original statement of the brachistochrone problem said nothing about functions $x(t)$. However, for a point to move via gravity from $A$ to $B$ in the shortest time, it is clear that it must follow the graph of a continuous function. Our shorthand notation for this property of the function $x(t)$ is $x \in C^0$. We also use the shorthand notation $x \in C^1$ to indicate that $x(t)$ is differentiable with continuous derivative $\dot{x}(t)$, and $x \in C^2$ to indicate that $x(t)$ is twice-differentiable with continuous first and second derivatives $\dot{x}(t)$ and $\ddot{x}(t)$:

$$x \in C^0 \iff x(t) \text{ is continuous,}$$
$$x \in C^1 \iff x(t) \text{ and } \dot{x}(t) \text{ are continuous,}$$
$$x \in C^2 \iff x(t), \dot{x}(t) \text{ and } \ddot{x}(t) \text{ are continuous.}$$

Notice that our statement of the brachistochrone problem via the integral functional $J[x]$ defined in (1.1) implicitly assumes that we can compute the corresponding derivatives $\dot{x}(t)$. Since we can always break the integral at a finite number of discontinuities, we only need the derivatives $\dot{x}(t)$ to be *piecewise continuous* on $[0,1]$, which means that there are at most a finite number of inputs $t \in [0,1]$ at which the derivative function $\dot{x}(t)$ is not continuous. This leads us to one last shorthand notation:

$$x \in \mathcal{D}^1 \iff x(t) \text{ is continuous, and } \dot{x}(t) \text{ is piecewise continuous.}$$

It follows that our version of the brachistochrone problem can be stated compactly as

$$\text{Minimize } J[x] = \int_0^1 \sqrt{\frac{1+\dot{x}^2}{1-x}}\, dt \quad \text{over } x \in \mathcal{D}^1 \text{ on } [0,1] \tag{1.2}$$

$$\text{with } x(0) = 1 \text{ and } x(1) = 0.$$

The admissible functions in (1.2) are the $x \in \mathcal{D}^1$ that satisfy the endpoint conditions $x(0) = 1$ and $x(1) = 0$.

**Problem 1.2.**

(a) *It should be clear that the function-categories satisfy the relationships*

$$\mathcal{C}^2 \subseteq \mathcal{C}^1 \subseteq \mathcal{C}^0,$$

*so that, for instance, a function $x \in \mathcal{C}^2$ is necessarily in the other two categories. How does the category $\mathcal{D}^1$ fit in this scheme?*

(b) *Sketch the graph of one example of a $\mathcal{D}^1$ function that is not a $\mathcal{C}^1$ function?*

### 1.1.3 Solution

To solve the brachistochrone problem (1.2), we will learn how to minimize integral functionals. The variables in this case are functions $x(t)$ themselves, and in the following chapter, we will see how to construct "variations" of a function to use for comparison in the minimization. These objects give the subject "calculus of variations" part of its name, and their development was motivated by the brachistochrone problem and other problems like it proposed in that era. We can use calculus of variations to show in particular that the solution to the brachistochrone problem (1.2) is a piece of a "cycloid". Figure 1.3 illustrates how a cycloid can be generated by rolling a circle from left to right along a horizontal axis.

**Restricted admissibility [3, adapted from Chapter 1]**

To develop our intuition for a solution to the brachistochrone problem, we will first restrict our attention to the family of admissible functions $x_\epsilon(t) = 1 - t^\epsilon$ for $\epsilon > 0$, some of whose graphs are shown in Figure 1.4. If we restrict the brachistochrone problem to this set of admissible functions $x_\epsilon(t)$, then we get

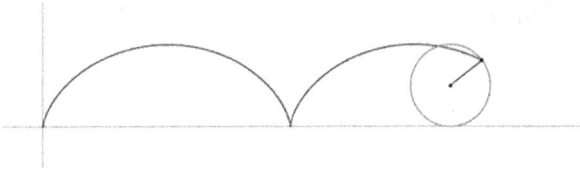

**Figure 1.3:** Cycloid generated by rolling circle.

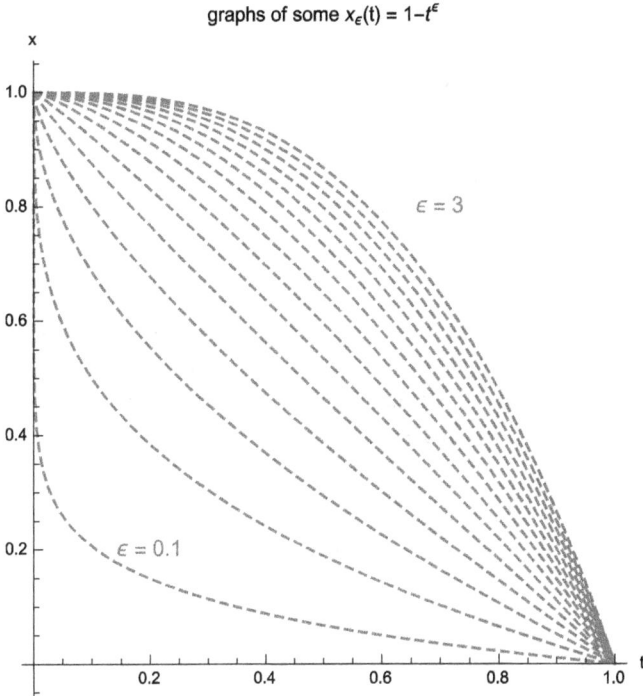

graphs of some $x_\epsilon(t) = 1 - t^\epsilon$

$\epsilon = 3$

$\epsilon = 0.1$

**Figure 1.4:** Admissible functions $x_\epsilon(t) = 1 - t^\epsilon$.

$$\text{Minimize } J[x_\epsilon] = \int_0^1 \sqrt{\frac{1 + \dot{x}_\epsilon^2}{1 - x_\epsilon}}\, dt = \int_0^1 t^{-\frac{\epsilon}{2}}\sqrt{1 + \epsilon^2 t^{2\epsilon - 2}}\, dt \qquad (1.3)$$

over $\epsilon > 0$. This minimization is the usual kind in terms of the single variable (number) $\epsilon$, and we can see the minimizing $\epsilon \approx 0.54$ from the graph in Figure 1.5, which was generated by Mathematica via the code

```
Plot[NIntegrate[t^(-ep/2) Sqrt[1+ep^2 t^(2 ep-2)],{t,0,1}],{
   ep,0.1,1}]
```

The minimum value satisfies $J[x_{0.54}] \approx 2.586$, which is clearly less than the value $J[x_1] = 2\sqrt{2} \approx 2.828$ we obtained in Problem 1.1(b) for the straight-line function $x_1(t) = 1 - t$, but

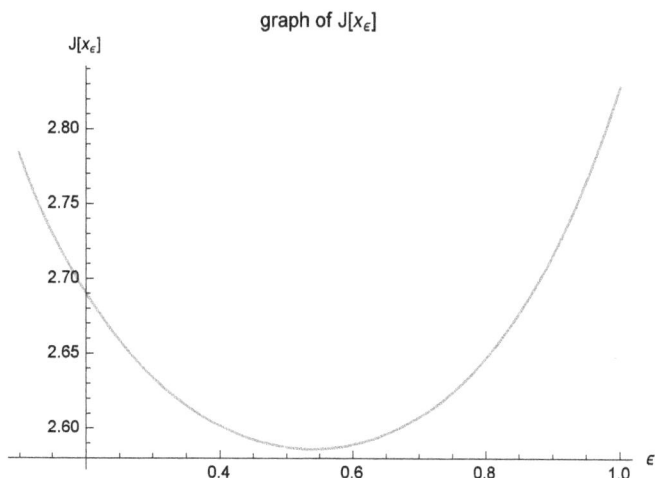

**Figure 1.5:** Graph of $J[x_\epsilon]$.

is the function $x_{0.54} = 1 - t^{0.54}$ the solution to the brachistochrone problem? It turns out that the answer is no. As we already noted, the true minimizing function is a piece of a cycloid, which is most easily expressed in parametric form (we will use the parameter $s$ and the superscript star $*$ to signal that this is a solution):

$$x^*(s) = 1 - \left(\frac{\cos(s)}{0.9342}\right)^2$$

$$t^*(s) = \frac{s + \sin(s)\cos(s) + \frac{\pi}{2}}{0.8727} \qquad \text{for } s \in \left[-\frac{\pi}{2}, 0.9342\right].$$

The word "parameter" is meant to signal a special kind of variable on which the other variables depend. Figure 1.6 shows the graphs of $x^*$ and $x_{0.54}$, where we can observe that these functions are different. The minimum value of the integral functional (1.1) is 2.582 (achieved by $x^*$), which is slightly less than the minimum value $J[x_{0.54}] \approx 2.586$ we obtained above for the minimization in (1.3).

## 1.2 Ancient history?

Even though the calculus of variations was first developed several hundred years ago, it has remained an active area of study. One notable more recent researcher was J. Ernest Wilkins, Jr., pictured in Figure 1.7, who completed a PhD at age 19 from the University of Chicago with a thesis titled *Multiple Integral Problems in Parametric Form in the Calculus of Variations*. He also has a one-page paper titled *The Silverman Necessary Conditions for Multiple Integrals in the Calculus of Variations* published in the Proceedings of the American Mathematical Society.

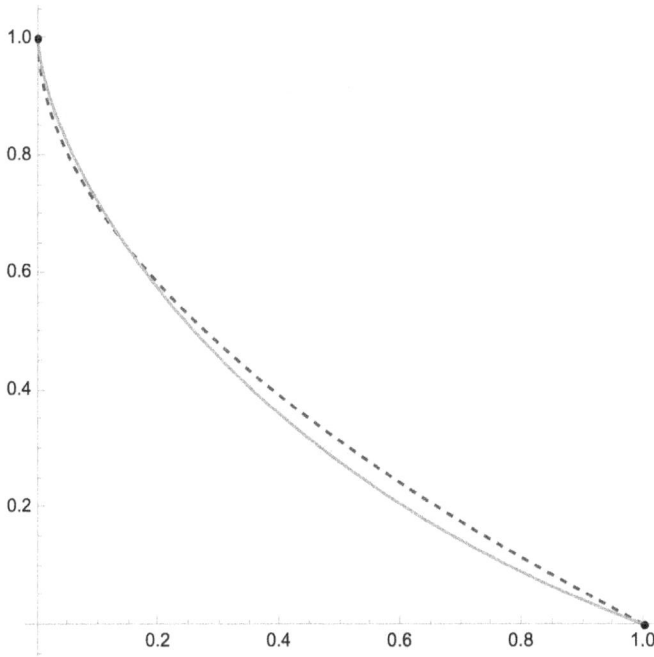

**Figure 1.6:** Solution to brachistochrone problem versus $x_{0.54}$ (dashed).

**Figure 1.7:** J. Ernest Wilkins, Jr.[1]

## Exercises

1-1. [3, adapted Exercise 1.2] Consider the calculus of variations problem

$$\text{Minimize } J[x] = \int_0^1 x^2 \dot{x}^2 \, dt \quad \text{over } x \in \mathcal{D}^1 \text{ on } [0,1]$$

$$\text{with } x(0) = 0 \text{ and } x(1) = 1. \tag{1.4}$$

a) Explain why the functions $x_\epsilon(t) = t^\epsilon$ parameterized by $\epsilon > 0.25$ are admissible.

b) Estimate the $J[x_\epsilon]$-minimizing parameter $\epsilon^*$ in this case.

c) Compute $J[x_{\epsilon^*}]$ and, without solving for the minimum value of the original problem (1.4), explain how $J[x_{\epsilon^*}]$ must compare to it.

1-2. Consider a function $x(t)$ satisfying $x(0) = 1$ and $x(1) = 2$. The curve formed by the piece of the graph of $x(t)$ connecting the points $(0, 1)$ and $(1, 2)$ in the $tx$-plane can be rotated about the $t$-axis to form a surface of revolution whose surface area is given by

$$J[x] = 2\pi \int_0^1 x \sqrt{1 + \dot{x}^2} \, dt.$$

a) Identify an appropriate (simple) family of admissible functions $x_\epsilon(t)$ parameterized by $\epsilon > 0$.

b) Estimate the $J[x_\epsilon]$-minimizing parameter $\epsilon^*$ in this case and give the surface area associated with $x_{\epsilon^*}$.

1-3. Write the calculus of variations problem for the shortest path between two points $(a, b)$ and $(c, d)$.

(Hint: Google "arclength".)

# 2 Euler–Lagrange equation

In the preceding chapter, we introduced an example of a calculus of variations problem where an integral functional $J[x]$ was minimized over a set of admissible functions $x(t)$. Now we will carefully consider a general "fixed-endpoint" calculus of variations problem of the form

$$\text{Minimize } J[x] = \int_{t_0}^{t_1} f(t, x, \dot{x}) \, dt \quad \text{over } x \in C^2 \text{ on } [t_0, t_1]$$

$$\text{with } x(t_0) = x_0 \text{ and } x(t_1) = x_1, \tag{2.1}$$

where the admissible functions $x$ are twice-continuously differentiable on the interval $[t_0, t_1]$, and the integral functional $J[x]$ is defined by a general *integrand* function $f$ of three arguments $(t, x, \dot{x})$. For example, the integrand function in the brachistochrone problem was

$$f(t, x, \dot{x}) = \sqrt{\frac{1 + \dot{x}^2}{1 - x}},$$

which did not happen to depend directly on the argument $t$.

We learn in calculus that to minimize a (continuously differentiable) function, we can identify the inputs that generate a zero derivative with respect to the input. This approach is challenging in the context of integral functionals since it requires us to "take a derivative" of the integral functional $J[x]$ with respect to input functions $x(t)$. To define such a derivative, we would need to compare distances between input functions (in the denominator of a difference-quotient), which is beyond the scope of this text.

Instead of developing notions of distances between functions here, we will follow the more direct approach developed by the Swiss mathematician and physicist Leonhard Euler and his successor as director of mathematics at the Prussian Academy of Sciences, the Italian-born mathematician and astronomer Joseph-Louis Lagrange.

## 2.1 Variations

The key object in this approach is a *variation* $x(t)$ from a presumed minimizing (admissible) function $x^*(t)$:

$$x(t) = x^*(t) + \epsilon \eta(t), \tag{2.2}$$

defined in terms of a small number $\epsilon$ (not necessarily positive) and a function $\eta(t)$ with $\eta \in C^2$ on $[t_0, t_1]$ satisfying $\eta(t_0) = \eta(t_1) = 0$. Figure 2.1 shows graphs of some variations

https://doi.org/10.1515/9783111290157-002

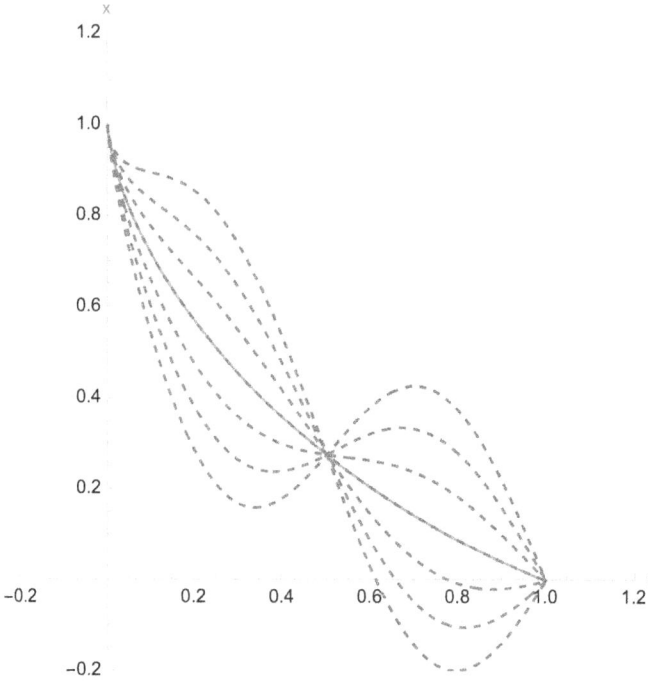

**Figure 2.1:** Variations $x(t) = x^*(t) + \epsilon \sin(2\pi t)$.

of the form $x(t) = x^*(t) + \epsilon \sin(2\pi t)$ from the minimizer of the brachistochrone problem, where the different curves are generated by different values of $\epsilon$.

**Question 2.1.** *How do we know that the variations $x(t) = x^*(t) + \epsilon\eta(t)$ are admissible for problem (2.1)?*

## 2.2 Derivation

If $x^*(t)$ is indeed a minimizer of the calculus of variations problem (2.1), then we know that it certainly satisfies $J[x^*] \le J[x]$ for any variation $x(t)$. This inequality immediately leads to

$$0 \le J[x] - J[x^*] = \int_{t_0}^{t_1} f(t, x, \dot{x}) - f(t, x^*, \dot{x}^*) \, dt, \tag{2.3}$$

where we have combined the two integrand functions in one integral, since the integrals used the same endpoints of integration.

To carry this further, we fix $t$ and approximate $f(t, \cdot, \cdot)$ with a first-order Taylor series in its second two arguments (i. e., a tangent plane) about the pair $(x^*, \dot{x}^*)$:

$$f(t, x, \dot{x}) = \underbrace{f(t, x^*, \dot{x}^*) + (x - x^*)f_x(t, x^*, \dot{x}^*) + (\dot{x} - \dot{x}^*)f_{\dot{x}}(t, x^*, \dot{x}^*)}_{\text{tangent plane for } f(t, \cdot, \cdot)} + \cdots$$

$$= f(t, x^*, \dot{x}^*) + \epsilon \eta f_x(t, x^*, \dot{x}^*) + \epsilon \dot{\eta} f_{\dot{x}}(t, x^*, \dot{x}^*) + O(\epsilon^2). \tag{2.4}$$

Note that we have applied (2.2) to deduce that

$$x - x^* = \epsilon \eta \quad \text{and} \quad \dot{x} - \dot{x}^* = \epsilon \dot{\eta},$$

and the notation $O(\epsilon^2)$ means that $\epsilon^2$ is the lowest power of $\epsilon$ possible in the rest of the series:

$$O(\epsilon^2) = c_2 \epsilon^2 + c_3 \epsilon^3 + \cdots$$

for some coefficients $c_i$. (Note that the Taylor series representation of $f(t, x, \dot{x})$ in (2.4) is valid as long as $f(t, \cdot, \cdot)$ has continuous first derivatives and bounded second derivatives.)

**Time out for Taylor series**

In calculus of one-variable functions $f(x)$, we use the following formula for the Taylor series of $f(x)$ about $x = a$:

$$f(a) + f'(a)(x - a) + f''(a)\frac{(x - a)^2}{2} + \cdots,$$

where the dots at the end mean that we keep adding terms of the form $f^{(i)}(a)\frac{(x-a)^i}{i!}$ forever. If the difference $(x - a)$ happens to be a multiple of $\epsilon$, then we can write the Taylor series as

$$f(a) + f'(a)(x - a) + O(\epsilon^2),$$

since the terms with a second derivative or higher are constant multiples of $(x - a)^i$ for $i \geq 2$.

The Taylor series of a function of two variables $f(x, y)$ about $(x, y) = (a, b)$ looks like

$$f(a, b) + f_x(a, b)(x - a) + f_y(a, b)(y - b) + \frac{f_{xx}(a, b)}{2}(x - a)^2$$

$$+ f_{xy}(a, b)(x - a)(y - b) + \frac{f_{yy}(a, b)}{2}(y - b)^2 + \cdots,$$

where there are now two first-derivative terms and three second-derivative terms (and even more higher-derivative terms to come). Now, if the differences $(x - a)$ and $(y - b)$ both happen to be multiples of $\epsilon$, then we can write the Taylor series as

$$f(a, b) + f_x(a, b)(x - a) + f_y(a, b)(y - b) + O(\epsilon^2),$$

since the terms with second derivatives or higher are constant multiples of $(x-a)^i(y-b)^j$ for $(i + j) \geq 2$. This is exactly what is happening in (2.4).

Substituting the Taylor series representation (2.4) into inequality (2.3), we get (suppressing the arguments of the two partial derivative functions for simplicity)

$$0 \leq \int_{t_0}^{t_1} \epsilon \eta f_x + \epsilon \dot{\eta} f_{\dot{x}} + O(\epsilon^2) \, dt$$

$$= \epsilon \int_{t_0}^{t_1} (\eta f_x + \dot{\eta} f_{\dot{x}}) \, dt + O(\epsilon^2)$$

$$= \epsilon \int_{t_0}^{t_1} \left( \eta f_x - \eta \frac{d}{dt}(f_{\dot{x}}) \right) dt + O(\epsilon^2) \tag{2.5}$$

$$= \epsilon \int_{t_0}^{t_1} \eta \left( f_x - \frac{d}{dt}(f_{\dot{x}}) \right) dt + O(\epsilon^2), \tag{2.6}$$

where we used integration by parts to get equation (2.5).

Dividing both sides of inequality (2.6) by $\epsilon > 0$ and then taking the limit as $\epsilon \to^+ 0$, we get

$$0 \leq \lim_{\epsilon \to^+ 0} \left( \int_{t_0}^{t_1} \eta \left( f_x - \frac{d}{dt}(f_{\dot{x}}) \right) dt + O(\epsilon) \right) = \int_{t_0}^{t_1} \eta \left( f_x - \frac{d}{dt}(f_{\dot{x}}) \right) dt.$$

If we instead divide (2.6) by $\epsilon < 0$ and take the limit as $\epsilon \to^- 0$, then we get the opposite inequality

$$0 \geq \lim_{\epsilon \to^- 0} \left( \int_{t_0}^{t_1} \eta \left( f_x - \frac{d}{dt}(f_{\dot{x}}) \right) dt + O(\epsilon) \right) = \int_{t_0}^{t_1} \eta \left( f_x - \frac{d}{dt}(f_{\dot{x}}) \right) dt,$$

since dividing by negative $\epsilon < 0$ reverses the inequality. We conclude that the equation

$$0 = \int_{t_0}^{t_1} \eta(t) \underbrace{\left( f_x(t, x^*, \dot{x}^*) - \frac{d}{dt}(f_{\dot{x}}(t, x^*, \dot{x}^*)) \right)}_{f_x - \frac{d}{dt}(f_{\dot{x}})} dt$$

holds for all functions $\eta \in C^2$ on $[t_0, t_1]$ satisfying $\eta(t_0) = \eta(t_1) = 0$.

**Question 2.2.** *What can we deduce from this about the term*

$$f_x(t, x^*, \dot{x}^*) - \frac{d}{dt}(f_{\dot{x}}(t, x^*, \dot{x}^*))?$$

This motivates the famous "Euler–Lagrange" differential equation identifying "extremals" for the calculus of variations problem (2.1).

**Euler–Lagrange Equation**

$$f_x(t, x^*, \dot{x}^*) - \frac{d}{dt}(f_{\dot{x}}(t, x^*, \dot{x}^*)) = 0$$

Any function $x^* \in C^2$ on $[t_0, t_1]$ satisfying the Euler–Lagrange equation is an *extremal* for the calculus of variations problem (2.1). Notice that the minimizer $x^*$ is an extremal but that the same argument works if we start with a maximizer, which is therefore also an extremal. Thus the Euler–Lagrange equation identifies a category of solution candidates for the minimization problem (2.1), and this category may contain other functions (including the maximizers).

**Applying the Euler–Lagrange equation**

The partial derivatives $f_x$ and $f_{\dot{x}}$ are computed symbolically first, without regard to any connection between the variables $x$ and $\dot{x}$. For example, if the integrand function is

$$f(t, x, \dot{x}) = x^2 \dot{x}^3,$$

then we get $f_x = 2x\dot{x}^3$ and $f_{\dot{x}} = 3x^2\dot{x}^2$ by considering the symbols $x$ and $\dot{x}$ as if they are entirely independent variables.

Then dependence on $t$ is identified everywhere it exists before the final derivative with respect to $t$ is taken. In our example here, the Euler–Lagrange equation translates as

$$\underbrace{2x(t)\dot{x}(t)^3}_{f_x(t, x(t), \dot{x}(t))} - \frac{d}{dt}\underbrace{(3x(t)^2\dot{x}(t)^2)}_{f_{\dot{x}}(t, x(t), \dot{x}(t))} = 0$$

$$\Downarrow$$

$$2x(t)\dot{x}(t)^3 - (6x(t)\dot{x}(t)\dot{x}(t)^2 + 6x(t)^2\dot{x}(t)\ddot{x}(t)) = 0,$$

where we used the product and chain rules to compute the derivative with respect to $t$. We typically rearrange the resulting differential equation so that $\ddot{x}(t)$ is isolated on one side and the equation is simplified:

$$\ddot{x}(t) = \frac{2x(t)\dot{x}(t)^3 - 6x(t)\dot{x}(t)\dot{x}(t)^2}{6x(t)^2\dot{x}(t)} = -\frac{2\dot{x}(t)^2}{3x(t)},$$

and then we solve the differential equation (if possible) to identify extremals.

**Problem 2.1.** *Find an extremal satisfying the endpoint conditions of*

$$Minimize \int_1^2 (t\dot{x})^2 \, dt \quad over \ x \in C^2 \ on \ [1,2]$$

$$with \ x(1) = 0 \ and \ x(2) = 1.$$

## 2.3 Solving the Euler–Lagrange equation

A variety of different application software solve differential equations, and we will provide coding suggestions for both Python and Mathematica.

**Python**
Python has a built-in dsolve command for solving differential equations like the Euler–Lagrange equation. When dsolve is able to solve a particular Euler–Lagrange equation, we can graph an extremal with the plot command.

For example, we can solve the Euler–Lagrange equation from Problem 2.1 via the Python commands

```
#import commands and define variables
from sympy import symbols, Function, diff, dsolve, Eq, plot
t = symbols("t")
x = Function('x')

#solve the Euler-Lagrange equation
print(dsolve(Eq(diff(x(t),t,2),-2*diff(x(t),t)/t)))
```

to obtain

```
Eq(x(t), C1 + C2/t)
```

which translates to the family of extremals $x^*(t) = c_1 + \frac{c_2}{t}$ defined by arbitrary constants $c_1$ and $c_2$. Then we enforce the endpoint conditions $x(1) = 0$ and $x(2) = 1$ to find the particular constants $c_1 = 2$ and $c_2 = -2$ associated with the extremal we seek, and we can graph that extremal on the domain of interest via the command

```
plot(2-2/t,(t,1,2),xlabel='t',ylabel='x(t)')
```

**Mathematica**

Mathematica has a built-in DSolve command for solving differential equations like the Euler–Lagrange equation. When DSolve is able to solve a particular Euler–Lagrange equation, we can graph the resulting extremal with the Plot command.

For example, we can solve the Euler–Lagrange equation from Problem 2.1 via the commands

```
DSolve[x''[t] = = -2 x'[t]/t, x[t], t]
```

to obtain

```
x[t] -> -C[1]/t + C[2]
```

which translates to the family of extremals $x^*(t) = -\frac{c_1}{t} + c_2$ defined by arbitrary constants $c_1$ and $c_2$. We determine the arbitrary constants as before by enforcing the endpoint conditions and then can graph that extremal on the domain of interest via the command

```
Plot[2-2/t, {t, 1, 2}, AxesLabel -> {"t", "x(t)"}]
```

**Failure?**

When the built-in commands fail to solve a particular Euler–Lagrange equation, we may still visualize an extremal by guessing an initial slope and using the "shooting method" (more on this in Section 2.3.1). For instance, here is the Mathematica code for applying the shooting method in Problem 2.1 using an initial slope guess of $x'(1) = 1$:

```
Plot[x[s] /. NDSolve[ {x''[t] = = -2 x'[t]/t, x[1] = = 0, x
    [2] = = 1}, x, {t, 1, 2},
  Method -> {"Shooting", "StartingInitialConditions" -> {x
    [1] = = 0, x'[1] = = 1}}],
  {s, 1, 2}, AxesLabel -> {"t", "x(t)"}]
```

### 2.3.1 Shooting method

The idea behind the shooting method can be understood by recalling a trick for converting a single second-order differential equation into a pair of first-order differential equations. First, we (algebraically) solve for $\ddot{x}$ after working out the Euler–Lagrange equation: $\ddot{x} = EL(t, x, \dot{x})$. For example, in Problem 2.1 the function $EL$ is just

$$EL(t, x, \dot{x}) = \frac{-2\dot{x}}{t}.$$

Then we introduce a dummy variable $v$ for the first derivative of $x$:

$$\dot{x} = v,$$
$$\dot{v} = EL(t, x, v),$$

getting an equivalent pair of first-order differential equations. Finally, we apply a numerical method (like Euler's method) to this pair of first-order differential equations.

To make the numerical method work, we need initial values for both $x$ and $v = \dot{x}$. Since our endpoint conditions only give us the first of these $x(t_0) = x_0$, the shooting method samples a variety of initial slopes $\dot{x}(t_0)$ until the value of $x$ at $t_1$ equals the target value $x(t_1) = x_1$ from the second endpoint condition. Essentially, the method "shoots" different-sloped solution trajectories from the initial endpoint condition until the second endpoint condition "target" is hit. Figure 2.2 shows the result of the shooting method applied via the Mathematica code above to the Euler–Lagrange equation from Problem 2.1, and here is a Python version of the same code:

```
#import commands
from numpy import dot, array
from matplotlib.pyplot import plot
from scipy.integrate import solve_ivp
from scipy.optimize import fsolve


#define EL and endpoint conditions
def EL(t,x,xdot): return -2*xdot/t
t0 = 1
t1 = 2
```

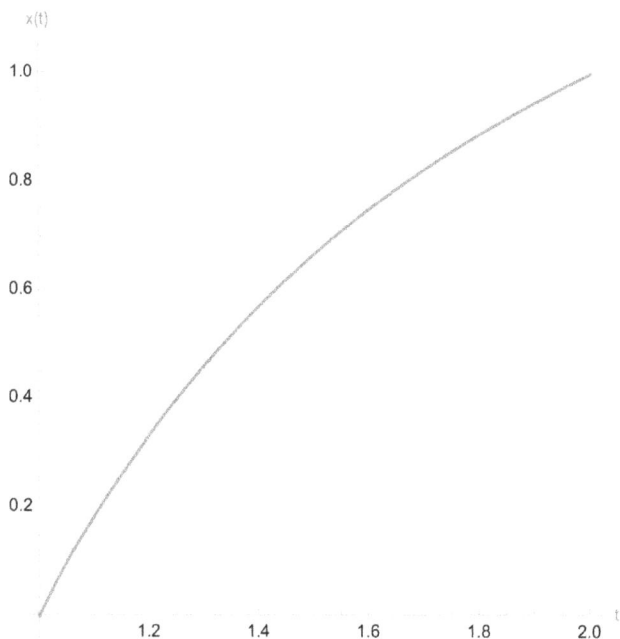

**Figure 2.2:** Shooting Method for Problem 2.1.

```
x0 = 0
x1 = 1

#shooting method
def desys(t, x): return dot(array([[0,1],[0,EL(t,x[0],x[1])
    ]]),x)
def objective(v0):
    sol = solve_ivp(desys, [t0,t1], [x0, v0])
    y = sol.y[0]
    return y[-1] - x1
v0 = fsolve(objective,1)
sol = solve_ivp(desys, [t0,t1], [x0, v0])
plot(sol.t, sol.y[0])
```

**Lab Activity 1.**

1.  *Graph an extremal satisfying the endpoint conditions of*

$$\text{Minimize } \int_0^1 x^2 \dot{x}^2 \, dt \quad \text{over } x \in C^0 \text{ on } [0,1]$$

$$\text{with } x(0) = 0 \text{ and } x(1) = 1.$$

2.  *Graph an extremal satisfying the endpoint conditions of*

$$\text{Minimize } 2\pi \int_0^1 x \sqrt{1 + \dot{x}^2} \, dt \quad \text{over } x \in C^0 \text{ on } [0,1]$$

$$\text{with } x(0) = 1 \text{ and } x(1) = 2.$$

## Exercises

2-1. Use integration by parts to show that

$$\int_{t_0}^{t_1} \dot{\eta} f_{\dot{x}} \, dt = - \int_{t_0}^{t_1} \eta \frac{d}{dt} (f_{\dot{x}}) \, dt.$$

2-2. Find an extremal satisfying the endpoint conditions of

$$\text{Minimize } J[x] = \int_0^3 (\dot{x}^2 - 1)^2 \, dt \quad \text{over } x \in C^2 \text{ on } [0,3]$$

$$\text{with } x(0) = 1 \text{ and } x(3) = 2$$

and evaluate the integral at the extremal.

2-3. Find an extremal satisfying the endpoint conditions of

$$\text{Minimize} \int_0^2 \frac{\dot{x}^2}{2} + x\dot{x} + x + \dot{x}\, dt \quad \text{over } x \in C^2 \text{ on } [0,2]$$

with $x(0) = 1$ and $x(2) = 7$.

2-4. Give an argument showing that if the Euler–Lagrange equation holds, then the *integral Euler–Lagrange equation* holds too:

$$\frac{d}{dt}(f - \dot{x}f_{\dot{x}}) = f_t. \tag{2.7}$$

(Hint: symbolically differentiate the left-side of (2.7) with respect to $t$ and remember to use the chain rule on

$$f = f(t, x(t), \dot{x}(t)),$$

since it depends on $t$ in all three arguments.)

2-5. Use the integral Euler–Lagrange equation (2.7) to produce first-order differential equations

$$\dot{x} = \cdots$$

for the extremals associated with each of Exercises 2-2 and 2-3.

(Hint: avoid computing the derivative with respect to $t$ on the left side of (2.7).)

# 3 Dido and transversality

An ancient calculus of variations problem involves the story of the founding of Carthage by Phoenicians fleeing their home city Tyre with Princess Dido [4, Chapter 1]. Figure 3.1 shows what the ruins of Carthage looked like more recently. According to the story, when Princess Dido landed on the northern coast of Africa with her followers, she negotiated with a local warlord to purchase as much land as she could enclose with a bull's hide. Dido had the bull's hide turned into rope so that it could enclose more land, and then she essentially solved the calculus of variations problem of enclosing the most land with the rope. She used the (essentially straight) shoreline to her advantage and walked the rope inland from one point on the shore, then back to another point on the shore as in Figure 3.2. Dido got the most area for Carthage by following a semicircular path with appropriate radius

$$r = \frac{\text{length of rope}}{\pi}$$

(since the arclength of a semicircle is $\pi r$). Dido likely used geometric intuition to solve this problem, but now we will consider a calculus of variations formulation of the same problem.

**Figure 3.1:** Carthage.[1]

---

1 Image Credit: Patrick Verdier (https://commons.wikimedia.org/wiki/File:Ruines_de_Carthage.jpg)

https://doi.org/10.1515/9783111290157-003

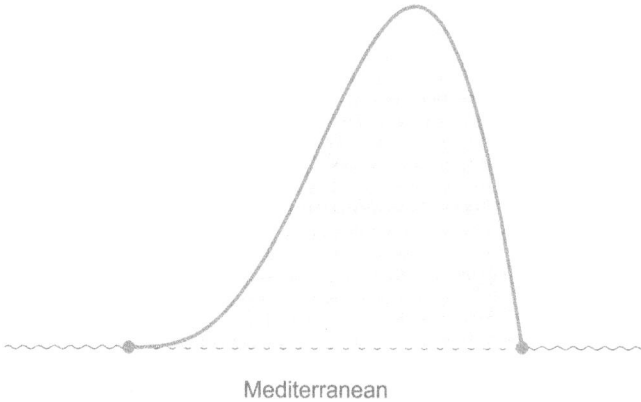

Figure 3.2: A possible boundary for New Carthage.

## 3.1 Dido's calculus of variations problem

Assuming that her starting point is at the origin of the $tx$-plane, Dido's problem was to choose a path-function $x(t)$ to follow inland, as well as a distance $T$ along the shore to end. Figure 3.3 shows the same image as in Figure 3.2, but with superimposed axes. It is clear from Figure 3.3 that the area of the land enclosed by the rope along this path is equal to the integral from 0 to $T$ of the path-function $x(t)$, so Dido's problem is closely related to the calculus of variations problem

$$\text{Minimize } J[x] = \int_0^T -x\,dt \quad \text{over } x \in C^0 \text{ on } [0, T]$$

$$\text{with } x(0) = 0 \text{ and } x(T) = 0, \tag{3.1}$$

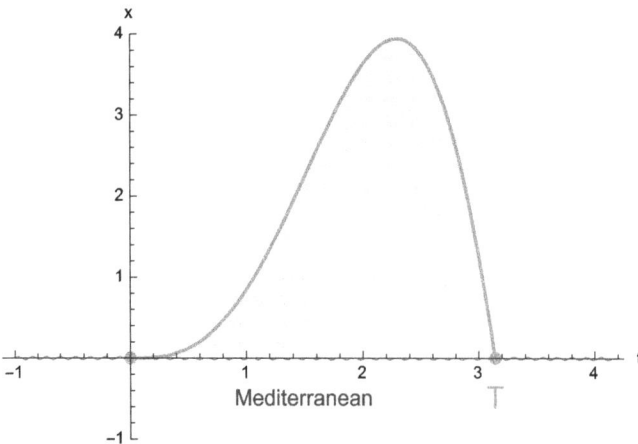

Figure 3.3: A possible boundary with superimposed axes.

where we minimize the negative area under the graph of $x(t)$ to maximize the area under the graph of $x(t)$.

**Question 3.1.** *Does the calculus of variations problem* (3.1) *have a solution?*

**Question 3.2.** *What restriction on Dido's problem have we so far neglected in our formulation* (3.1)?

We will return to the neglected restriction later, but now we will focus our attention on how the endpoint condition $x(T) = 0$ in this case is actually related to a "target curve".

## 3.2 Target curves

The endpoint conditions $x(0) = 0$ and $x(T) = 0$ in problem (3.1) are an example of a more general class of endpoint conditions

$$x(t_0) = x_0 \quad \text{and} \quad x(T) = \text{tar}(T)$$

where $T$ is a variable time, and $\text{tar}(t)$ is a given function whose graph represents a *target curve* as in Figure 3.4. Notice that different graphs of $x(t)$ in Figure 3.4 can intersect the target curve $\text{tar}(t)$ at different "end" times $T$, which is why $T$ is generally variable in these conditions.

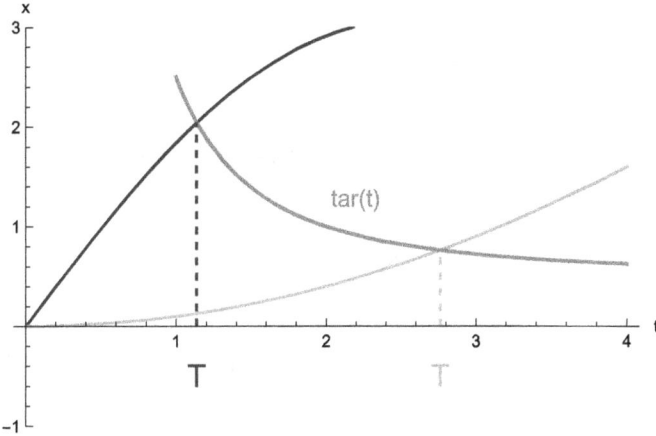

**Figure 3.4:** Target curve.

**Question 3.3.** *What does the target curve look like if $T = t_1$ is a fixed value, but $x(T)$ is free to take any value?*

The general "target curve" problem in the calculus of variations is

$$\text{Minimize } J[x] = \int_{t_0}^{T} f(t, x, \dot{x}) \, dt \quad \text{over } x \in C^2 \text{ on } [t_0, T]$$

$$\text{with } x(t_0) = x_0 \text{ and } x(T) = \text{tar}(T). \tag{3.2}$$

## 3.3 Transversality condition

In mathematics, *transversality* describes intersections and can be thought of as a companion to the notion of tangency. In a calculus of variations problem with a target curve, the admissible functions all intersect the target curve. As the following derivation shows, this intersection results in a "transversality condition", which must be satisfied by an extremal in this case.

As in our derivation of the Euler–Lagrange equations, we suppose we have a solution $x^*(t)$ to the general target curve problem (3.2). We let $T^*$ be the end time when the solution $x^*$ hits the target curve $x^*(T^*) = \text{tar}(T^*)$, and, as before, we compare $x^*(t)$ to an admissible variation $x(t) = x^*(t) + \epsilon\eta(t)$. In order for $x(T)$ to be admissible, the function $\eta \in C^2$ must in particular satisfy $\eta(t_0) = 0$ (so that $x(t_0) = x_0$ as before) and

$$\eta(T) = \frac{\text{tar}(T) - x^*(T)}{\epsilon} \tag{3.3}$$

for some end time $T$ (so that $x(T) = \text{tar}(T)$). (We will assume for now that $T \neq T^*$, since otherwise we would have $x(T^*) = x^*(T^*)$, which gives us the fixed-endpoint problem again.)

An important fact (which follows from $x^*$ and $x$ both being in $C^2$) is that the difference $\Delta T$ between the two different end times is $O(\epsilon)$:

$$\Delta T = T - T^* = O(\epsilon). \tag{3.4}$$

This quantifies the notion that as the variation $x(t)$ becomes closer to the solution $x^*(t)$ (by shrinking $\epsilon$), the respective end times $T$ and $T^*$ when their graphs intersect the target curve will also become closer.

### Comparing the integrals
The two integrals $J[x]$ and $J[x^*]$ to be compared have different end times, but we can align part of their difference as follows:

$$\Delta J = J[x] - J[x^*] = \int_{t_0}^{T} f(t, x, \dot{x}) \, dt - \int_{t_0}^{T^*} f(t, x^*, \dot{x}^*) \, dt$$

$$= \int_{t_0}^{T^*} \underbrace{f(t,x,\dot{x}) - f(t,x^*,\dot{x}^*)\,dt}_{A} + \underbrace{\int_{T^*}^{T} f(t,x,\dot{x})\,dt}_{B}. \qquad (3.5)$$

To work out the "left-over" integral $B$, we can use a Taylor series of the integrand function $f$ about $(T^*,x^*(T^*),\dot{x}^*(T^*))$:

$$f(t,x(t),\dot{x}(t)) = f(T^*,x^*(T^*),\dot{x}^*(T^*)) + O(\epsilon),$$

where we have applied relationship (3.4) to replace each difference $t - T^*$ in the Taylor series with an $O(\epsilon)$ (assuming that $t$ is between $T$ and $T^*$, we know that the difference $t - T^*$ is smaller in magnitude than $\Delta T = T - T^*$ and so inherits at least its $O(\epsilon)$ relationship). This leads immediately to the following evaluation of the integral $B$:

$$\int_{T^*}^{T} f(t,x,\dot{x})\,dt = \int_{T^*}^{T} f(T^*,x^*(T^*),\dot{x}^*(T^*)) + O(\epsilon)\,dt$$

$$= f(T^*)\Delta T + O(\epsilon^2), \qquad (3.6)$$

where we have applied (3.4) and taken advantage of the fact that the constant

$$f(T^*) = f(T^*,x^*(T^*),\dot{x}^*(T^*))$$

integrates very simply.

Exercise 3-4 confirms that the "aligned" integral $A$ satisfies

$$\int_{t_0}^{T^*} f(t,x,\dot{x}) - f(t,x^*,\dot{x}^*)\,dt = \epsilon\eta(T^*)\cdot f_{\dot{x}}(T^*) + O(\epsilon^2), \qquad (3.7)$$

where $f_{\dot{x}}(T^*) = f_{\dot{x}}(T^*,x^*(T^*),\dot{x}^*(T^*))$. Since the function $\eta(t)$ is largely mysterious, the term $\epsilon\eta(T^*)$ appearing in (3.7) is also mysterious. To make progress, we now develop a formula for this term that will allow us to remove the mystery from expression (3.7).

### A formula for $\epsilon\eta(T^*)$

We use relationship (3.4) to deduce that

$$O((T - T^*)^2) = O(\epsilon^2)$$

in the following two Taylor series about $T^*$:

$$x(T) = \underbrace{x^*(T^*) + \epsilon\eta(T^*)}_{x(T^*)} + \underbrace{(\dot{x}^*(T^*) + \epsilon\dot{\eta}(T^*))}_{\dot{x}(T^*)}\Delta T + O(\epsilon^2),$$

$$tar(T) = tar(T^*) + tar'(T^*)\Delta T + O(\epsilon^2).$$

We then substitute these equations into the "hits-the-target" equation $x(T) = \text{tar}(T)$ and solve for $e\eta(T^*)$:

$$\begin{aligned} e\eta(T^*) &= \text{tar}(T^*) - x^*(T^*) + (\text{tar}'(T^*) - \dot{x}^*(T^*) - e\dot{\eta}(T^*))\Delta T + O(\epsilon^2) \\ &= \text{tar}(T^*) - x^*(T^*) + (\text{tar}'(T^*) - \dot{x}^*(T^*))\Delta T + O(\epsilon^2) \\ &= (\text{tar}'(T^*) - \dot{x}^*(T^*))\Delta T + O(\epsilon^2), \end{aligned} \tag{3.8}$$

where we absorbed $e\dot{\eta}(T^*)\Delta T$ into the $O(\epsilon^2)$ term to get the second line via relationship (3.4), and we applied the "hits-the-target" equation $x^*(T^*) = \text{tar}(T^*)$ to get the third line.

### 3.3.1 The final transversality condition

We finally have all the pieces we need to construct a useful comparison of the two integrals $J[x]$ and $J[x^*]$ by substituting (3.6), (3.7), and (3.8) into expression (3.5):

$$\Delta J = (f(T^*) + (\text{tar}'(T^*) - \dot{x}^*(T^*))f_{\dot{x}}(T^*))\Delta T + O(\epsilon^2).$$

As in the derivation of the Euler–Lagrange equation, we can divide both sides of the inequality $0 \leq \Delta J$ (which holds since $x^*$ minimizes the integral) alternately by $\epsilon > 0$ and $\epsilon < 0$, and then take the limit as $\epsilon \to 0$. Relationship (3.4) then allows us to conclude the following:

**Transversality Condition**

$$f(T^*) + (\text{tar}'(T^*) - \dot{x}^*(T^*))f_{\dot{x}}(T^*) = 0$$

**Question 3.4.** *What does the transversality condition reduce to in the case where $x(T) = x_1$ is fixed, but $T$ is free to be anything?*

**Question 3.5.** *What does the transversality condition reduce to in the case where $x(T)$ is free, but $T = t_1$ is fixed?*
*(Hint: divide by $\text{tar}'(T^*)$ first.)*

### 3.3.2 Transversality and integral Euler–Lagrange

You may have noticed that the transversality condition and the integral Euler–Lagrange equation

$$\frac{d}{dt}(f - \dot{x}f_{\dot{x}}) = f_t$$

have some similar components. In the particular case where $f_t = 0$, these similarities can be exploited very conveniently, since then we know that the extremal $x^*$ satisfies the following simplified version of integral Euler–Lagrange:

$$f(t) - \dot{x}^*(t)f_{\dot{x}}(t) = \text{constant.} \tag{3.9}$$

Notice that we can rearrange the transversality condition as

$$f(T^*) - \dot{x}^*(T^*)f_{\dot{x}}(T^*) = -\text{tar}'(T^*)f_{\dot{x}}(T^*),$$

which means that the constant on the right side of (3.9) is evidently $-\text{tar}'(T^*)f_{\dot{x}}(T^*)$:

$$f(t) - \dot{x}^*(t)f_{\dot{x}}(t) = -\text{tar}'(T^*)f_{\dot{x}}(T^*).$$

In particular, when $\text{tar}'(T^*) = 0$ (and $f_t = 0$), the integral Euler–Lagrange equation reduces to

$$f(t) - \dot{x}^*(t)f_{\dot{x}}(t) = 0.$$

### 3.3.3 Using transversality

Notice that the transversality condition is an algebraic (not differential) equation. We can use this equation in a similar manner to an endpoint condition, along with the Euler–Lagrange (differential) equation to help us find extremals satisfying the endpoint conditions of general target curve problems of the form (3.2). The typical approach is as follows:

1. Identify extremals $x^*$ by solving the Euler–Lagrange equation.
2. Use the endpoint conditions to determine arbitrary constants.
3. Use the transversality condition to determine $T^*$ and any remaining arbitrary constants.

**Problem 3.1.** *Find an extremal satisfying the endpoint conditions of*

$$\text{Minimize } \int_1^T (\dot{x}t)^2 \, dt \quad \text{over } x \in C^2 \text{ on } [1, T]$$

$$\text{with } x(1) = 0 \text{ and } x(T) = \frac{1}{T^2}.$$

## Exercises

3-1. Find an extremal satisfying the endpoint conditions of

$$\text{Minimize } \int_1^T \frac{\dot{x}^2}{t^3}\, dt \quad \text{over } x \in C^2 \text{ on } [1, T]$$

$$\text{with } x(1) = -1 \text{ and } x(T) = (T - 1)^2$$

and find the optimal end time $T^*$.

3-2. Find an extremal satisfying the endpoint condition of

$$\text{Minimize } \int_0^\pi \dot{x}^2 + 2x \sin(t)\, dt \quad \text{over } x \in C^2 \text{ on } [0, \pi]$$

$$\text{with } x(0) = 0.$$

3-3. Find an extremal satisfying the endpoint conditions of

$$\text{Minimize } \int_0^T (\dot{x}^2 - 1)^2\, dt \quad \text{over } x \in C^2 \text{ on } [0, T]$$

$$\text{with } x(0) = 1 \text{ and } x(T) = 2$$

and find the optimal end time $T^*$.

3-4. Show that if $x^*$ is an extremal for problem (3.2), then

$$\int_{t_0}^{T^*} f(t, x, \dot{x}) - f(t, x^*, \dot{x}^*)\, dt = \epsilon\eta(T^*) \cdot f_{\dot{x}}(T^*) + O(\epsilon^2),$$

where $f_{\dot{x}}(T^*) = f_{\dot{x}}(T^*, x^*(T^*), \dot{x}^*(T^*))$.

(Hint: mimic the derivation of the Euler–Lagrange equation.)

# 4 Extremals versus minimizers

We have seen already that extremals are only candidates for the solution to a minimization problem, and now we will explore one approach that may allow us to confirm whether or not an extremal is a minimizer.

## 4.1 Directly verifying a minimizer

The most straightforward way to confirm that an extremal $x^*(t)$ is a minimizer is to compare it directly to all other admissible functions $x(t)$. It will be useful to express any such admissible function as $x(t) = x^*(t) + \eta(t)$, where $\eta$ is simply the difference $\eta(t) = x(t) - x^*(t)$. This construction is similar to our earlier one for variations $x(t) = x^*(t) + \epsilon\eta(t)$, but now we do not use the $\epsilon$.

**Example 1.** Consider the calculus of variations problem

$$\text{Minimize } \int_1^2 (\dot{x}t)^2 \, dt \quad \text{over } x \in C^2 \text{ on } [1,2]$$

$$\text{with } x(1) = 0 \text{ and } x(2) = 1. \tag{4.1}$$

We have already found the extremal $x^*(t) = 2 - \frac{2}{t}$ for this problem, and we know that any other admissible function $x(t)$ can be written as $x(t) = x^*(t) + \eta(t)$, where $\eta(t) \in C^2$ and satisfies $\eta(1) = \eta(2) = 0$ (since $x$ satisfies the same endpoint conditions as $x^*$).

To see whether $x^*(t)$ is a minimizer of $J$, we can compare its value $J[x^*]$ directly to $J[x]$:

$$J[x] - J[x^*] = \int_1^2 \left( \underbrace{\frac{2}{t^2} + \dot{\eta}}_{\dot{x}} \right)^2 t^2 - \left( \underbrace{\frac{2}{t^2}}_{\dot{x}^*} \right)^2 t^2 \, dt$$

$$= \int_1^2 4\dot{\eta} \, dt + \int_1^2 \dot{\eta}^2 t^2 \, dt,$$

$$(\text{FTC}) \rightarrow = 4\eta(t) \Big|_1^2 + \int_1^2 \dot{\eta}^2 t^2 \, dt,$$

$$(\eta(1) = \eta(2) = 0) \rightarrow = \int_1^2 \dot{\eta}^2 t^2 \, dt$$

$$\geq 0,$$

where FTC stands for the fundamental theorem of calculus, and the final inequality follows since the integrand satisfies $\dot{\eta}^2 t^2 \geq 0$ on $[1,2]$. We conclude that $J[x] \geq J[x^*]$, which means that $x^*$ is a minimizer among all admissible functions.

https://doi.org/10.1515/9783111290157-004

Unfortunately, this "direct" method works relatively rarely (when manipulations allow us to work out an inequality in this way).

**Problem 4.1.** *Find the extremal $x^*$ satisfying the endpoint conditions of*

$$\text{Minimize } \int_0^3 (\dot{x}^2 - 1)^2 \, dt \quad over \ x \in C^2 \ on \ [0,3]$$

$$with \ x(0) = 1 \ and \ x(3) = 2 \tag{4.2}$$

*and directly compare $J[x^*]$ to $J[x]$.*


## 4.2 Considering $\mathcal{D}^1$

There are many $\mathcal{D}^1$ functions connecting the endpoints $(0,1)$ and $(3,2)$ in the $tx$-plane and which evaluate to zero in the integral $\int_0^3 (\dot{x}^2 - 1)^2 \, dt$ from (4.2).

**Question 4.1.** *What does the graph of one such function look like?*

You should see that each piece separately satisfies the Euler–Lagrange equation. Any such $\mathcal{D}^1$ function solves the calculus of variations problem if we expand our category of admissible functions to be $\mathcal{D}^1$:

$$\text{Minimize } \int_0^3 (\dot{x}^2 - 1)^2 \, dt \quad over \ x \in \mathcal{D}^1 \ on \ [0,3]$$

$$with \ x(0) = 1 \ and \ x(3) = 2. \tag{4.3}$$

**Question 4.2.** *How can you be sure that the $\mathcal{D}^1$ function you graphed in Question 4.1 minimizes the calculus of variations problem (4.3)?*

**Question 4.3.** *How could we construct a $C^2$ function with nearly the same integral value as such a $\mathcal{D}^1$ function? (Hint: think graphically.)*

**Question 4.4.** *What does this imply about the existence of a $C^2$ minimizer to (4.2)?*

**Problem 4.2.** *Explain why the (verified) $C^2$ minimizer from Example 1 must still be a minimizer if we expand the admissible functions to $\mathcal{D}^1$.*


**Conclusions: $\mathcal{D}^1$ versus $C^2$**
- If $x^*$ is a minimizer in $C^2$, then it is also a minimizer in $\mathcal{D}^1$.
- There may be a minimizer in $\mathcal{D}^1$ even if there is no minimizer in $C^2$.

The same statements are true with $C^1$ in place of $C^2$.

**Lab Activity 2.**

1. *Find an extremal for*

$$\text{Minimize} \int_0^2 x^2(\dot{x} - 1)^2 \, dt \quad \text{over } x \in C^2 \text{ on } [0, 2]$$

*satisfying $x(0) = 0$ and $x(2) = d$ (if one exists) when*
(a) $d = 2$
(b) $d = 1$
(c) $d = 0$

2. *Find minimizers of*

$$\int_0^2 x^2(\dot{x} - 1)^2 \, dt \quad \text{over } x \in \mathcal{D}^1 \text{ on } [0, 2]$$

*satisfying each pair of endpoint conditions given in part 1.*

## Exercises

In each case below, find a minimizer if one exists. Justify your result thoroughly and be sure to give the corresponding minimum value. If no minimizer exists, then thoroughly justify that statement.

1. Minimize $\int_0^1 (\dot{x} - 1)^2 \, dt$ over $x \in \mathcal{D}^1$ on $[0, 1]$ with $x(0) = 0$ and $x(1) = 2$.
2. Minimize $\int_1^2 \frac{\dot{x}^2}{t^3} \, dt$ over $x \in C^2$ on $[1, 2]$ with $x(1) = -1$ and $x(2) = 4$.
3. Minimize $\int_0^2 (\dot{x}^2 - 4)^2 \, dt$ over $x \in \mathcal{D}^1$ on $[0, 2]$ with $x(0) = 1$ and $x(2) = 0$.
4. Minimize $\int_0^2 (\dot{x}^2 - 4)^2 \, dt$ over $x \in C^2$ on $[0, 2]$ with $x(0) = 1$ and $x(2) = 0$.
5. Minimize $\int_0^2 (\dot{x} - 1)^2 \, dt$ over $x \in \mathcal{D}^1$ on $[0, 2]$ with $x(0) = 1$ and $x(2) = 0$.

# 5 Second-derivative test

In the preceding chapter, we used direct verification to determine when an extremal is
a minimizer in some situations. We also saw that direct verification has limitations, so
now we will develop another way of investigating whether an extremal is a minimizer.

The Euler–Lagrange equation is analogous to the first-derivative test from calculus
(i. e., the first derivative is zero at a minimizer), and serves the same role in identifying a
pool of candidate minimizers. In calculus, we use a second-derivative test to narrow this
pool of candidates, and now we will develop a "second-derivative test" to do the same
thing in the calculus of variations.

## 5.1 Derivation

In Chapter 1, we used a first-order Taylor approximation to derive the Euler–Lagrange
equation for the fixed-endpoint calculus of variations problem

$$\text{Minimize } \underbrace{\int_{t_0}^{t_1} f(t, x, \dot{x}) \, dt}_{J[x]} \quad \text{over } x \in C^2 \text{ on } [t_0, t_1]$$

$$\text{with } x(t_0) = x_0 \text{ and } x(t_1) = x_1. \tag{5.1}$$

Now we will follow a similar derivation but with second-order terms included explicitly
in the Taylor approximation.

Supposing that $x^*(t)$ is a minimizer of (5.1), we know as before that any variations
$x = x^* + \epsilon\eta$ for $\eta \in C^2$ satisfying $\eta(t_0) = \eta(t_1) = 0$ have the relationship

$$0 \le \Delta J = J[x] - J[x^*] = \int_{t_0}^{t_1} f(t, x, \dot{x}) - f(t, x^*, \dot{x}^*) \, dt.$$

We can then substitute a second-order Taylor approximation for $f(t, \cdot, \cdot)$ to conclude that

$$0 \le \Delta J = \epsilon V_1 + \epsilon^2 V_2 + O(\epsilon^3), \tag{5.2}$$

where

$$V_1 = \int_{t_0}^{t_1} \eta f_x + \dot{\eta} f_{\dot{x}} \, dt \quad \text{(first-order terms)},$$

$$V_2 = \frac{1}{2} \int_{t_0}^{t_1} \eta^2 f_{xx} + 2\eta\dot{\eta} f_{x\dot{x}} + \dot{\eta}^2 f_{\dot{x}\dot{x}} \, dt \quad \text{(second-order terms)}.$$

https://doi.org/10.1515/9783111290157-005

(Note that in our derivation of the Euler–Lagrange equation from Chapter 2, the terms $\epsilon^2 V_2 + O(\epsilon^3)$ were combined in a single term $O(\epsilon^2)$.) Separately dividing both sides of (5.2) by $\epsilon > 0$ and $\epsilon < 0$ and taking the limit in each case as $\epsilon \to 0$ give $V_1 = 0$ as before. It follows that we can drop $V_1$ from (5.2) to get

$$0 \leq \Delta J = \epsilon^2 V_2 + O(\epsilon^3).$$

Dividing both sides by $\epsilon^2$ and taking the limit as $\epsilon \to 0$ yield the inequality

$$\underbrace{\frac{1}{2} \int_{t_0}^{t_1} \eta^2 f_{xx} + 2\eta\dot{\eta} f_{x\dot{x}} + \dot{\eta}^2 f_{\dot{x}\dot{x}} \, dt}_{V_2} \geq 0. \tag{5.3}$$

To simplify this, we can apply integration by parts to the middle term in the integral with $u = f_{x\dot{x}}$ and $\frac{dv}{dt} = 2\eta\dot{\eta}$:

$$\int_{t_0}^{t_1} 2\eta\dot{\eta} f_{x\dot{x}} \, dt = \eta^2 f_{x\dot{x}} \Big]_{t_0}^{t_1} - \int_{t_0}^{t_1} \eta^2 \frac{d}{dt}(f_{x\dot{x}}) \, dt$$

$$(\eta(t_0) = \eta(t_1) = 0) \to = -\int_{t_0}^{t_1} \eta^2 \frac{d}{dt}(f_{x\dot{x}}) \, dt.$$

Substituting this into our condition (5.3) and factoring out the $\eta^2$ give

$$\frac{1}{2} \int_{t_0}^{t_1} \eta^2 \left( f_{xx} - \frac{d}{dt}(f_{x\dot{x}}) \right) + \dot{\eta}^2 f_{\dot{x}\dot{x}} \, dt \geq 0. \tag{5.4}$$

### 5.1.1 From $C^2$ to $\mathcal{D}^1$

In Chapter 4, we learned that since $x^*(t)$ is a (presumed) minimizer of (5.1), it is also a minimizer of the broader problem where admissibility is defined by $x \in \mathcal{D}^1$ in place of $x \in C^2$. Since the work above is unaffected by this change, inequality (5.4) actually holds for any variations constructed from $\eta \in \mathcal{D}^1$ (satisfying $\eta(t_0) = \eta(t_1) = 0$). From this we can deduce (see the proof to come) the following simpler condition due to the French mathematician Adrien-Marie Legendre:

**Second-Derivative Test**

$$x^* \text{ minimizer} \implies f_{\dot{x}\dot{x}}(t, x^*(t), \dot{x}^*(t)) \geq 0 \text{ for } t \in [t_0, t_1]$$

The practical purpose of any test like this is identifying calculations that can allow us to make conclusions about optimality. The implication in the second-derivative test means that we can *rule out* $x^*$ as a minimizer by computing the second derivative $f_{\dot{x}\dot{x}}$ and determining that it is negative for some $t$ in the interval $[t_0, t_1]$.

**Proof**

To simplify the notation, we define the (continuous) function of $t$

$$g(t) = f_{xx}(t, x^*(t), \dot{x}^*(t)) - \frac{d}{dt}(f_{x\dot{x}}(t, x^*(t), \dot{x}^*(t))),$$

in which case inequality (5.4) can be written as

$$\frac{1}{2}\int_{t_0}^{t_1} \eta^2 g + \dot{\eta}^2 f_{\dot{x}\dot{x}} \, dt \geq 0. \tag{5.5}$$

To show that a minimizer $x^*$ satisfies the inequality in the second-derivative test, we suppose that it does not [3, adapted from proof of Lemma 7.1] and look for a contradiction to inequality (5.5). By continuity we know that a negative value of $f_{\dot{x}\dot{x}}(t, x^*(t), \dot{x}^*(t))$ anywhere in the interval $[t_0, t_1]$ guarantees negative values (below some threshold $-a$ for $a > 0$) on a subinterval $I = [b - \beta, b + \beta] \subseteq [t_0, t_1]$:

$$f_{\dot{x}\dot{x}}(t, x^*(t), \dot{x}^*(t)) \leq -a \quad \text{on } I. \tag{5.6}$$

Now we define the $\mathcal{D}^1$ function

$$\eta(t) = \begin{cases} \sin^2(\frac{\pi(t-b)}{\beta}) & \text{if } t \in I, \\ 0 & \text{otherwise} \end{cases}$$

(for which a sample graph appears in Figure 5.1) and substitute it into inequality (5.5):

$$\frac{1}{2}\int_{b-\beta}^{b+\beta} \underbrace{\left(\sin^4\left(\frac{\pi(t-b)}{\beta}\right)\right)}_{\eta^2} g(t) + \underbrace{\frac{4\pi^2}{\beta^2}\sin^2\left(\frac{\pi(t-b)}{\beta}\right)\cos^2\left(\frac{\pi(t-b)}{\beta}\right)}_{\dot{\eta}^2} f_{\dot{x}\dot{x}} \, dt \geq 0. \tag{5.7}$$

Since $g$ is continuous, we can bound its value on $I = [b - \beta, b + \beta]$ by some $c > 0$. This can be used with (5.6) to deduce the following upper bound for the integral on the left side of (5.7):

$$\frac{1}{2}\int_{b-\beta}^{b+\beta} \left(\sin^4\left(\frac{\pi(t-b)}{\beta}\right)\right) g(t) + \frac{4\pi^2}{\beta^2}\sin^2\left(\frac{\pi(t-b)}{\beta}\right)\cos^2\left(\frac{\pi(t-b)}{\beta}\right) f_{\dot{x}\dot{x}} \, dt$$

$I = [.5, 1.5]$

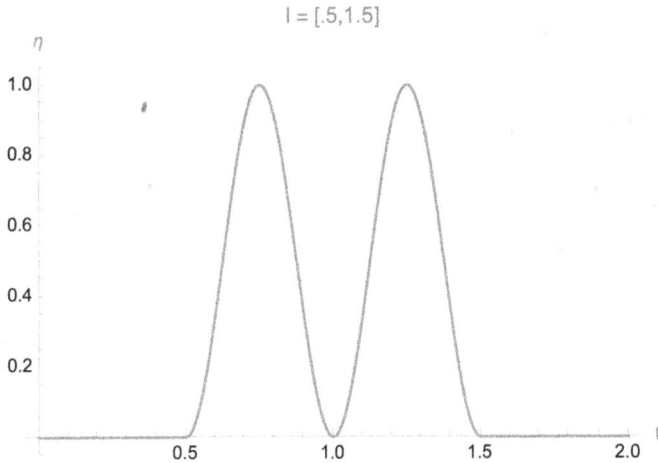

**Figure 5.1:** Graph of $\eta(t)$ for $b = 1$ and $\beta = 0.5$.

$$\leq \frac{1}{2} \int_{b-\beta}^{b+\beta} \left( \sin^4\left(\frac{\pi(t-b)}{\beta}\right)c + \frac{4\pi^2}{\beta^2} \sin^2\left(\frac{\pi(t-b)}{\beta}\right)\cos^2\left(\frac{\pi(t-b)}{\beta}\right)(-a) \right) dt$$

$$\leq \frac{c}{2} \int_{b-\beta}^{b+\beta} \left( \sin^4\left(\frac{\pi(t-b)}{\beta}\right) \right) dt - \frac{a}{2}\left(\frac{4\pi^2}{\beta^2}\right) \int_{b-\beta}^{b+\beta} \sin^2\left(\frac{\pi(t-b)}{\beta}\right)\cos^2\left(\frac{\pi(t-b)}{\beta}\right) dt$$

$$= \frac{c}{2}\left(\frac{3\beta}{4}\right) - \frac{a}{2}\left(\frac{4\pi^2}{\beta^2}\right)\left(\frac{\beta}{4}\right)$$

$$= \frac{3c\beta}{8} - \frac{\pi^2 a}{2\beta}. \tag{5.8}$$

If we shrink $\beta > 0$ arbitrarily close to zero, then we get the same bound (since the same argument applies). However, in this case the right side of (5.8) eventually is negative (since the values $a$ and $c$ need not change with $\beta$). This contradicts (5.7) (which is a rewritten inequality (5.5)), and we conclude that the second-derivative test is valid.

## 5.2 Applying the second-derivative test

**Problem 5.1.** *Consider the following calculus of variations problems:*

(a) *Minimize $\int_1^2 x^2 t^3 \, dt$ over $x \in C^2$ on $[1, 2]$ with $x(1) = 0$ and $x(2) = 3$,*

(b) *Minimize $\int_0^3 (\dot{x}^2 - 1)^2 \, dt$ over $x \in C^2$ on $[0, 3]$ with $x(0) = 1$ and $x(3) = 2$,*

(c) *Minimize $\int_0^1 \frac{1}{x} \, dt$ over $x \in C^2$ on $[0, 1]$ with $x(0) = 0$ and $x(1) = 1$,*

*with extremals $x^* = 4 - \frac{4}{t}, x^* = \frac{t}{3} + 1$, and $x^* = t$, respectively. What is the most we can conclude from the second-derivative test in each case?*

**Problem 5.2.** *For the calculus of variations problem*

$$\text{Minimize} \int_0^1 \frac{1}{\dot{x}} \, dt \quad \text{over } x \in C^2 \text{ on } [0,1]$$

$$\text{with } x(0) = 0 \text{ and } x(1) = 1:$$

(a) *Construct a function $x \in \mathcal{D}^1$ satisfying $J[x] < J[x^*]$ for the extremal $x^* = t$;*
(b) *Evaluate (using $x^* = t$ and $\eta = \frac{x-t}{\epsilon}$ with $x$ from part (a))*

$$V_1 = \int_{t_0}^{t_1} \eta f_x + \dot{\eta} f_{\dot{x}} \, dt,$$

$$V_2 = \frac{1}{2} \int_{t_0}^{t_1} \eta^2 f_{xx} + 2\eta\dot{\eta} f_{x\dot{x}} + \dot{\eta}^2 f_{\dot{x}\dot{x}} \, dt,$$

*and explain how*

$$\Delta J = \epsilon V_1 + \epsilon^2 V_2 + O(\epsilon^3)$$

*can be negative (as it must be since $J[x] < J[x^*]$).*

## Exercises

5-1. Without solving for an extremal, what is the most we can conclude from the second-derivative test for the brachistochrone problem

$$\text{Minimize} \int_0^1 \sqrt{\frac{1 + \dot{x}^2}{1 - x}} \, dt \quad \text{over } x \in C^2 \text{ on } [0,1]$$

$$\text{with } x(0) = 1 \text{ and } x(1) = 0.$$

Assume that admissible functions $x$ also satisfy $x(t) \le 1$ on $[0,1]$ so that the integrand function does not produce complex values.

5-2. [3, adapted Exercise 2.9] Suppose $x(t)$ is the production rate at $t$ (in years) for some factory and that $x \in C^2$. If manufacturing costs change at an annual rate of $c\dot{x}^2$ (for a constant cost parameter $c > 0$) and personnel costs change at the annual rate of $a c t \dot{x}$ (for a constant $a \in [0,1]$), then:

(a) Write the calculus of variations problem for the (total) cost-minimizing production rate if the initial rate is $x(0) = p$ and the factory wants to double the production rate in one year.
(b) Find an extremal $x^*$ satisfying the endpoint conditions of this problem.

(c) How high must the initial production rate $p$ be to guarantee that production rate $x^*$ always increases throughout the year?

(d) What is the most we can conclude from the second-derivative test in this case?

5-3. What is the most we can conclude from the second-derivative test for the calculus of variations problem

$$\text{Minimize} \int_0^2 (\dot{x}^2 - 2)^2 dt \quad \text{over } x \in C^2 \text{ on } [0, 2]$$

$$\text{with } x(0) = 0 \text{ and } x(2) = 1?$$

5-4. Find an extremal $x^*$ satisfying the endpoint conditions of

$$\text{Minimize} \int_0^2 x^2(\dot{x} - 1)^2 dt \quad \text{over } x \in C^2 \text{ on } [0, 2]$$

$$\text{with } x(0) = 0 \text{ and } x(2) = 1$$

and explain what the second-derivative test says about $x^*$ as a minimizer of this calculus of variations problem.

# 6 Strong second-derivative test

In the preceding chapter, we derived a calculus of variations analogue to the second-derivative test from calculus and saw that, like the second-derivative test from calculus, it is only "necessary" for optimality. That is, our second-derivative test is only *necessary* for an extremal $x^*$ to be a minimizer (i. e., $x^*$ must satisfy it) but not *sufficient* (i. e., $x^*$ might not be a minimizer even if it satisfies the second-derivative test). As a result, we can only use our second-derivative test to rule out minimizer candidates. Now we will develop a "stronger" second-derivative test, which is sufficient for optimality and so will allow us to rule *in* minimizing candidates. The derivation of this test will use an embedding of an extremal $x^*$ in a family of functions called a "field of extremals".

## 6.1 Fields of extremals

We know that extremals are solutions to the Euler–Lagrange equation, and we have seen that these functions typically include two arbitrary constants. If we enforce a single endpoint condition, then we will get a family of functions parameterized by a single arbitrary constant. This family is one possibility for a "field of extremals".

**Example 2.** The calculus of variations problem

$$\text{Minimize } \int_0^1 \dot{x}^3 \, dt \quad \text{over } x \in \mathcal{C}^2 \text{ on } [0,1]$$

$$\text{with } x(0) = 0 \text{ and } x(1) = 1$$

has the Euler–Lagrange equation $6\dot{x}\ddot{x} = 0$ with solutions $x(t) = at + b$. If we enforce the first endpoint condition $x(0) = 0$, then we get the family of functions $x(t) = at$ parameterized by the single arbitrary constant $a$. By graphing this family of functions as in Figure 6.1, we can see how it "covers" the $tx$-plane.

We say that a particular point $(\bar{t}, \bar{x})$ in the $tx$-plane is *covered* by a function $x(t)$ if the graph of $x$ passes through the point: $x(\bar{t}) = \bar{x}$.

**Question 6.1.** *What points in the tx-plane are not covered by any member of the family of functions $x(t) = at$ from Example 2?*

**Question 6.2.** *Are there any points in the tx-plane covered by more than one member of the family of functions $x(t) = at$ from Example 2?*

We say that a family of functions *simply covers* a region $R$ of the $tx$-plane if the graph of exactly one function in the family passes through each pair $(t, x) \in R$. For instance, the family of functions $x(t) = at$ shown in Figure 6.1 simply covers the $tx$-plane minus the $x$-axis (the origin is excluded too since it is covered more than once).

https://doi.org/10.1515/9783111290157-006

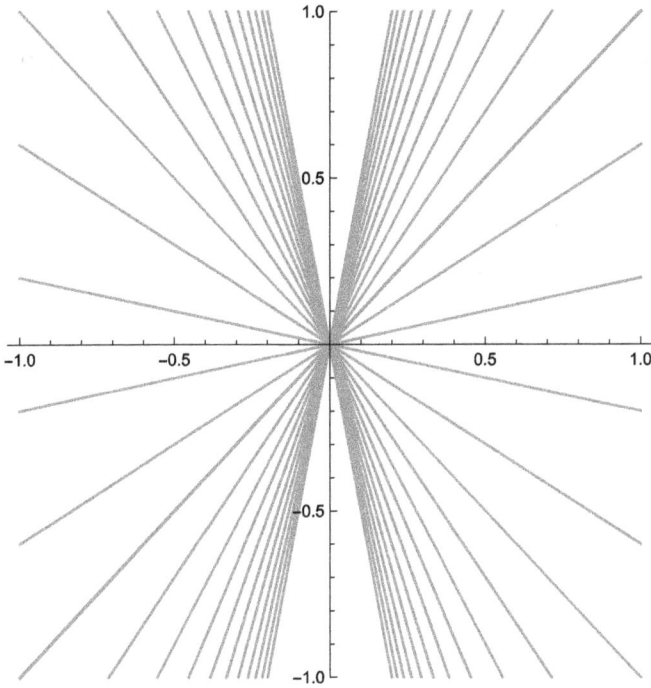

**Figure 6.1:** Graphs $x(t) = at$ for various $a$.

Given a calculus of variations problem and a region $R$ of the $tx$-plane, a parameterized family of functions $\mathcal{F}$ is a *field of extremals for R* if

- Every function in $\mathcal{F}$ is an extremal (i. e., satisfies the Euler–Lagrange equation);
- $\mathcal{F}$ includes an extremal $x^*(t)$ satisfying the endpoint conditions, and
- $\mathcal{F}$ simply covers $R$.

**Problem 6.1.** *For the calculus of variations problem*

$$Minimize \int_0^1 \dot{x}^3\, dt \quad over\ x \in C^2\ on\ [0,1]$$

$$with\ x(0) = 0\ and\ x(1) = 1,$$

*find a field of extremals for the entire tx-plane.*

## 6.2 Sufficient condition

Associated with any field of extremals $\mathcal{F}$ is a *slope function* $p(t, x) = \dot{x}(t)$ giving the slope of the (unique) function $x(t) \in \mathcal{F}$ through the point $(t, x)$ in the (simply covered) region $R$.

**Question 6.3.** *What is the slope function corresponding to the field of extremals $\mathcal{F} = \{t + b : b \in \mathbb{R}\}$ for the entire tx-plane? What is the slope function corresponding to the field of extremals $\mathcal{F} = \{at : a \in \mathbb{R}\}$ for the tx-plane minus the x-axis?*

### 6.2.1 A related integral

We consider again the general fixed-endpoint problem from (2.1):

$$\text{Minimize } J[x] = \int_{t_0}^{t_1} f(t, x, \dot{x}) \, dt \quad \text{over } x \in C^2 \text{ on } [t_0, t_1]$$

$$\text{with } x(t_0) = x_0 \text{ and } x(t_1) = x_1$$

and find a field of extremals $\mathcal{F}$ that simply covers the *region of integration R*, which is the section of the *tx*-plane directly above and below the (open) interval $(t_0, t_1)$ on the *t*-axis:

$$R = \{\text{all pairs } (t, x) \text{ with } t \in (t_0, t_1)\},$$

as illustrated in Figure 6.2. Since the region of integration excludes the vertical lines through $t_0$ and $t_1$, we will usually be able to construct a field of extremals simply by enforcing one endpoint condition and not the other. This works because the overcovering inherent in this construction always happens either along the vertical line through $t_0$ or the vertical line through $t_1$ (depending on which endpoint condition we enforce).

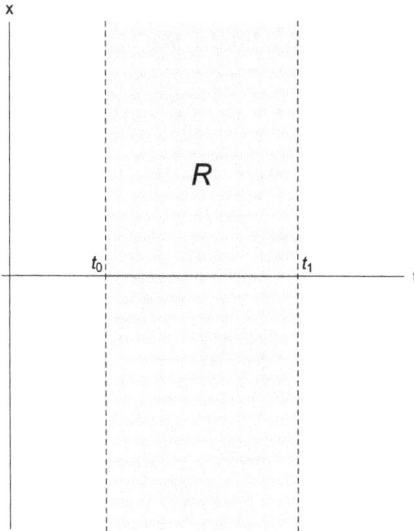

**Figure 6.2:** Region of integration for (2.1).

Then we construct a new integral functional $K[x]$ (sometimes called "Hilbert's invariant integral", after German mathematician David Hilbert) by integrating the tangent line at $p$ of the integrand function $f(t, x, \cdot)$:

$$K[x] = \int_{t_0}^{t_1} \overbrace{f(t,x,p) + (\dot{x} - p)f_{\dot{x}}(t,x,p)}^{\text{tangent line at } p \text{ of } f(t,x,\cdot)} \, dt, \qquad (6.1)$$

where we interpret $p$ as shorthand for the slope function $p(t, x(t))$ evaluated along the graph of the function $x(t)$ (which is not necessarily in the field of extremals $\mathcal{F}$).

Figure 6.3 shows the field of extremals $\mathcal{F} = \{t + b : b \in \mathbb{R}\}$, where the slope function is $p(t, x) = 1$. The dashed curve in Figure 6.3 represents the graph of some function $x(t)$

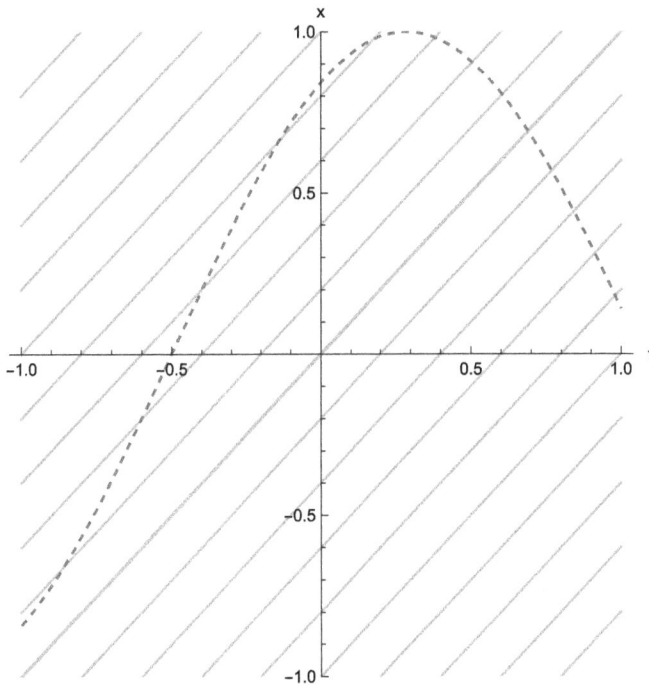

**Figure 6.3:** $\mathcal{F} = \{t + b : b \in \mathbb{R}\}$ with $x(t)$ not in $\mathcal{F}$.

not in $\mathcal{F}$, where it is clear that $\dot{x}(t) \neq 1$ at most locations. It follows that the integral (6.1) evaluated along this $x$ satisfies

$$K[x] = \int_{t_0}^{t_1} f(t, x, 1) + (\dot{x} - 1)f_{\dot{x}}(t, x, 1) \, dt,$$

where we have applied $p(t, x) = 1$ for this slope function. If we instead evaluate the integral (6.1) along any member $x(t) = t + b$ of the field of extremals, then we get

$$K[x] = \int_{t_0}^{t_1} f(t, x, \dot{x})\, dt = J[x],$$

because $\dot{x} = 1 = p$.

**Question 6.4.** *How is $K[x^*]$ related to $J[x^*]$ in general?*

### 6.2.2 Path independence

After defining the vector field

$$F(t, x) = \underbrace{(f(t, x, p) - pf_{\dot{x}}(t, x, p))}_{f_1(t,x)}\, \vec{\imath} + \underbrace{f_{\dot{x}}(t, x, p)}_{f_2(t,x)}\, \vec{\jmath} \tag{6.2}$$

in terms of the canonical unit vectors $\vec{\imath}$ and $\vec{\jmath}$ (in the $x_1$- and $x_2$-directions, respectively), where $p$ represents the given slope function $p(t, x)$, we can usefully rewrite the new integral functional from (6.1), evaluated at any admissible $x$, as the line integral

$$K[x] = \int_{t_0}^{t_1} f(t, x, p) + (\dot{x} - p)f_{\dot{x}}(t, x, p)\, dt$$

$$= \int_{t_0}^{t_1} F(r(t)) \cdot r'(t)\, dt = \int_C F \cdot dr$$

along the curve $C$ parameterized as $r(t) = t\vec{\imath} + x(t)\vec{\jmath}$. The vector field $F$ defined in (6.2) satisfies the curl test

$$\frac{\partial f_1(t, x)}{\partial x} = \frac{\partial f_2(t, x)}{\partial t},$$

so the integral $K[x]$ is path independent. Since $x$ connects the same endpoints as $x^*$, we conclude that $K[x^*] = K[x]$.

### 6.2.3 Strong second-derivative test

We now put all the pieces together to derive the following equation for $\Delta J$:

$$\Delta J = J[x] - J[x^*]$$

$$= J[x] - K[x^*] \quad (\text{since } K[x^*] = J[x^*])$$
$$= J[x] - K[x] \quad (\text{since } K[x^*] = K[x])$$
$$= \int_{t_0}^{t_1} \underbrace{f(t, x, \dot{x}) - f(t, x, p) - (\dot{x} - p)f_{\dot{x}}(t, x, p)}_{\text{exc}(t, x, \dot{x}, p)} \, dt \tag{6.3}$$

in terms of the *excess* function $\text{exc}(t, x, \dot{x}, p) = f(t, x, \dot{x}) - f(t, x, p) - (\dot{x} - p)f_{\dot{x}}(t, x, p)$. It follows immediately from (6.3) that if the excess is nonnegative,

$$\text{exc}(t, x, \dot{x}, p) \geq 0 \quad \text{for all } (x, \dot{x}) \text{ and } t \in [t_0, t_1], \tag{6.4}$$

then we can conclude that $\Delta J \geq 0$, since then there is nonnegative signed area under a nonnegative excess function graph. Since $\Delta J \geq 0$ guarantees that $x^*$ is a minimizer, we know that the nonnegative excess (6.4) guarantees the same.

You may recognize the excess function exc as the difference between the integrand function $f(t, x, \cdot)$ at $\dot{x}$ and its tangent line at $p$. The nonnegative excess (6.4) thus demands that the graph of the integrand function $f(t, x, \cdot)$ always lies above its tangent line at $p$, which ensures a kind of convexity similar to the second-derivative test from calculus. To turn the nonnegative excess into a more familiar kind of second-derivative test, we can use a second-order Taylor remainder term to rewrite the excess function as

$$\text{exc}(t, x, \dot{x}, p) = \frac{1}{2}(\dot{x} - p)^2 f_{\dot{x}\dot{x}}(t, x, \xi)$$

in terms of a mystery point $\xi$ between $\dot{x}$ and $p$. If we substitute this into the nonnegative excess condition (6.4), then we get the following form of our sufficient condition:

**Strong Second-Derivative Test**

The extremal $x^*$ is a minimizer if $f_{\dot{x}\dot{x}}(t, x, \xi) \geq 0$ for all $(x, \xi)$ and $t \in [t_0, t_1]$.

Recall that our earlier second-derivative test only requires nonnegativity along the extremal $x^*$:

$$f_{\dot{x}\dot{x}}(t, x^*(t), \dot{x}^*(t)) \geq 0 \quad \text{for all } t \in [t_0, t_1],$$

which is less demanding than the strong second-derivative test (hence the modifier "strong"). Note that the stronger requirement of the strong second-derivative test guarantees a stronger consequence that every extremal $x^*$ is a minimizer.

### 6.2.4 Using the strong second-derivative test

The proof connecting the strong second-derivative test to minimization hinges on the existence of a field of extremals, so we should identify one to use this test to conclude that we have a minimizer as follows:

1. Identify an extremal $x^*$ and a field of extremals that simply covers the region of integration.
2. Compute the second derivative $f_{\dot{x}\dot{x}}$ and determine if it passes the strong second-derivative test.

Recall that we can usually find a field of extremals that simply covers the region of integration by solving the Euler–Lagrange equation and enforcing only one endpoint condition. The following example demonstrates that this does not always work.

**Example 3.**

$$\text{Minimize } J[x] = \int_0^{2\pi} \dot{x}^2 - x^2 \, dt \quad \text{over } x \in C^2 \text{ on } [0, 2\pi]$$

$$\text{with } x(0) = 0 \text{ and } x(2\pi) = 0.$$

The Euler–Lagrange equation in this case reduces to $\ddot{x} = -x$, and enforcing either endpoint condition gives the family of extremals $a\sin(t)$ parameterized by $a \in \mathbb{R}$. This family does not simply cover the region of integration in this case, because every member of the family passes through the point $(\pi, 0)$ in the $tx$-plane, and no member of the family passes through any point of the form $(\pi, b)$ for $b \neq 0$.

**Problem 6.2.** *Find and confirm a minimizer of the calculus of variations problem*

$$\text{Minimize } \int_0^2 \dot{x}^2 \, dt \quad \text{over } x \in C^2 \text{ on } [0, 2]$$

$$\text{with } x(0) = -1 \text{ and } x(2) = 3$$

*by identifying a field of extremals and checking the strong second-derivative test. Also show that the excess function is nonnegative (6.4).*

### 6.2.5 Gap between tests

We have seen a second-derivative test and a strong second-derivative test, and now we will consider an example that satisfies the first test but not the second.

**Example 4.**

$$\text{Minimize } J[x] = \int_0^1 \dot{x}^2 - 4x\dot{x}^3 + 2t\dot{x}^4 \, dt \quad \text{over } x \in C^2 \text{ on } [0,1]$$

$$\text{with } x(0) = 0 \text{ and } x(1) = 0.$$

The Euler–Lagrange equation in this case reduces to

$$(12t\dot{x}^2 - 12x\dot{x} + 1)\ddot{x} = 0,$$

and so $x^*(t) = 0$ is an extremal satisfying both endpoint conditions (since $\ddot{x}^* = 0$). To confirm either the second-derivative test or the strong second-derivative test, we compute

$$f_{\dot{x}\dot{x}}(t,x,\dot{x}) = 24t\dot{x}^2 - 24x\dot{x} + 2$$

and notice that $f_{\dot{x}\dot{x}}(t,x^*,\dot{x}^*) = 2 \geq 0$. Thus the second-derivative test is satisfied in this case.

Notice that $\mathcal{F} = \{b : b \in \mathbb{R}\}$ is a field of extremals covering the entire $tx$-plane and containing the extremal $x^*(t) = 0$ in Example 4, so if the strong second-derivative test is satisfied, then we can deduce that $x^*$ is a minimizer. However, if we choose the pair $(x, \xi) = (1,1)$ (for instance), then we have

$$f_{\dot{x}\dot{x}}(t,x,\xi) = 24t - 22 < 0 \quad \text{for } t < \frac{11}{12}.$$

It follows that the strong second-derivative test is not satisfied in this case. Note that $x^*(t) = 0$ is not a minimizer in this case, so this example again demonstrates the limitation of our original second-derivative test, as well as the significance of its strengthened version.

## Exercises

6-1. Find and confirm a minimizer for the problem

$$\text{Minimize } \int_1^2 \dot{x} + t^2 \dot{x}^2 \, dt \quad \text{over } x \in C^2 \text{ on } [1,2]$$

$$\text{with } x(1) = 0 \text{ and } x(2) = 4.$$

6-2. Find and confirm a minimizer for the problem

$$\text{Minimize } \int_0^2 \frac{\dot{x}^2}{2} + x\dot{x} + x + \dot{x} \, dt \quad \text{over } x \in C^2 \text{ on } [0,2]$$

$$\text{with } x(0) = 0 \text{ and } x(2) = 4.$$

6-3. Find and confirm a minimizer for the problem

$$\text{Minimize} \int_1^2 \dot{x}^2 + \frac{2x^2}{t^2} \, dt \quad \text{over } x \in C^2 \text{ on } [1,2]$$

with $x(1) = 0$ and $x(2) = 3$.

6-4. Show that the integral functional $K[x]$ from (6.1) satisfies

$$K[x] = \int_{t_0}^{t_1} \mathbf{F}(\mathbf{r}(t)) \cdot \mathbf{r}'(t) \, dt$$

for the vector field $\mathbf{F}$ defined by (6.2) and the parameterized curve $\mathbf{r}(t) = t\vec{\imath} + x(t)\vec{\jmath}$.

6-5. Show that the vector field

$$\mathbf{F}(t,x) = \underbrace{\left(f(t,x,p) - pf_{\dot{x}}(t,x,p)\right)}_{f_1(t,x)}\vec{\imath} + \underbrace{f_{\dot{x}}(t,x,p)}_{f_2(t,x)}\vec{\jmath}$$

satisfies the curl test

$$\frac{\partial f_1(t,x)}{\partial x} - \frac{\partial f_2(t,x)}{\partial t} = 0.$$

Hints: recall that $p = p(t,x)$ depends on both $t$ and $x$, use the fact that any function $x(t) \in \mathcal{F}$ satisfies $\ddot{x} = p_t + pp_x$ (which we get by the chain rule on $\dot{x} = p(t,x)$), and expand the right side of the Euler–Lagrange equation $f_x = \frac{d}{dt}f_{\dot{x}}(t,x(t),\dot{x}(t))$, which is satisfied by any such $x(t) \in \mathcal{F}$.

# 7 Integral constraints

In this chapter, we revisit Dido's problem of maximizing the area enclosed by a rope, but now we explicitly include the constraint that the rope has fixed length. This "isoperimetric" (i. e., same perimeter) problem is an example of a constrained calculus of variations problem. We will see how to solve constrained calculus of variations problems like this via a simple adaptation of the idea of Lagrange multipliers from calculus.

### Dido's isoperimetric problem

Recall that Dido wanted to maximize the area along a beach and enclosed by a rope whose outline is given by the graph of $x \in C^2$ in the $tx$-plane.

**Problem 7.1.** *If the rope has a fixed length of 100, write the calculus of variations problem that Dido solved. (Hint: recall the arclength formula.)*

### General integral constraints

Since integral constraints need to be distinguished from the integral functional $J[x]$ to be minimized, we refer to the latter integral as the *cost functional* (and the integrand function $f(t, x, \dot{x})$ as the *cost integrand*). Then the *constraint integrand* is the function $g(t, x, \dot{x})$ inside the integral constraint

$$I[x] = \int_{t_0}^{T} g(t, x, \dot{x}) \, dt = c.$$

In Dido's problem, the cost integrand is $f(t, x, \dot{x}) = -x$, the constraint integrand is $g(t, x, \dot{x}) = \sqrt{1 + \dot{x}^2}$, and the constant $c = 100$. A simple way to deal with integral constraints like this in calculus of variations problems is to borrow the concept of Lagrange multipliers from calculus.

## 7.1 Lagrange multipliers from calculus

The setting here is minimizing a function $F(\epsilon_1, \epsilon_2)$ of two variables subject to the equation constraint $G(\epsilon_1, \epsilon_2) = c$, written in terms of another function $G(\epsilon_1, \epsilon_2)$ of two variables, and some constant $c$. In calculus, we learned that the constrained minimizers will be among the pairs $(\epsilon_1, \epsilon_2)$ satisfying the *Lagrange multiplier rule*: $\nabla L(\epsilon_1, \epsilon_2) = \mathbf{0}$, written here in terms of the *Lagrangian function*

$$L(\epsilon_1, \epsilon_2) = F(\epsilon_1, \epsilon_2) + \lambda G(\epsilon_1, \epsilon_2)$$

https://doi.org/10.1515/9783111290157-007

for some *Lagrange multiplier* $\lambda$. The point of this rule is that the constrained minimizers are among the stationary (i. e., zero-gradient generating) points of the Lagrangian function, so we can apply methods for finding stationary points (of $L$ in this case) to solve the original constrained problem.

## 7.2 Lagrange multipliers in calculus of variations

We can apply the Lagrange multiplier rule from calculus to a calculus of variations problem with integral constraint by constructing variations on a (constrained) minimizer $x^*$ of the form

$$x_{\epsilon_1,\epsilon_2}(t) = x^*(t) + \epsilon_1\eta_1(t) + \epsilon_2\eta_2(t)$$

with two variables $(\epsilon_1, \epsilon_2)$. As usual, we consider only $\eta_1$ and $\eta_2$ that ensure that the variations are admissible. The cost functional evaluated along any such variation is then a simple function of two variables $F(\epsilon_1, \epsilon_2) = J[x_{\epsilon_1,\epsilon_2}]$ that is minimized at the point $(\epsilon_1, \epsilon_2) = (0, 0)$ over all points $(\epsilon_1, \epsilon_2)$ satisfying the constraint $G(\epsilon_1, \epsilon_2) = c$, where the constraint function $G$ of two variables is defined as $G(\epsilon_1, \epsilon_2) = I[x_{\epsilon_1,\epsilon_2}]$.

As it is a constrained minimizer, the point $(0, 0)$ must satisfy the usual Lagrange multiplier rule $\nabla L(0, 0) = \mathbf{0}$ for the two-variable Lagrangian function

$$L(\epsilon_1, \epsilon_2) = J[x_{\epsilon_1,\epsilon_2}] + \lambda I[x_{\epsilon_1,\epsilon_2}].$$

After computing the two first partial derivatives

$$L_{\epsilon_1}(\epsilon_1, \epsilon_2) \quad \text{and} \quad L_{\epsilon_2}(\epsilon_1, \epsilon_2)$$

of this Lagrangian function, we can translate the Lagrange multiplier rule $\nabla L(0, 0) = \mathbf{0}$ in this case as

$$\underbrace{\int_{t_0}^{T} \eta_1(f_x(t, x^*, \dot{x}^*) + \lambda g_x(t, x^*, \dot{x}^*)) + \dot{\eta}_1(f_{\dot{x}}(t, x^*, \dot{x}^*) + \lambda g_{\dot{x}}(t, x^*, \dot{x}^*))\, dt = 0,}_{L_{\epsilon_1}(\epsilon_1, \epsilon_2)}$$

$$\underbrace{\int_{t_0}^{T} \eta_2(f_x(t, x^*, \dot{x}^*) + \lambda g_x(t, x^*, \dot{x}^*)) + \dot{\eta}_2(f_{\dot{x}}(t, x^*, \dot{x}^*) + \lambda g_{\dot{x}}(t, x^*, \dot{x}^*))\, dt = 0.}_{L_{\epsilon_2}(\epsilon_1, \epsilon_2)}$$

Using integration by parts as in the derivation of the Euler–Lagrange equation, these two equations translate into

$$\int_{t_0}^{T} \eta_1\left(f_x(t,x^*,\dot{x}^*) + \lambda g_x(t,x^*,\dot{x}^*) - \frac{d}{dt}(f_{\dot{x}}(t,x^*,\dot{x}^*) + \lambda g_{\dot{x}}(t,x^*,\dot{x}^*))\right) dt = 0,$$

$$\int_{t_0}^{T} \eta_2\left(f_x(t,x^*,\dot{x}^*) + \lambda g_x(t,x^*,\dot{x}^*) - \frac{d}{dt}(f_{\dot{x}}(t,x^*,\dot{x}^*) + \lambda g_{\dot{x}}(t,x^*,\dot{x}^*))\right) dt = 0.$$

Since these must hold for any appropriate functions $\eta_1(t)$ and $\eta_2(t)$, we can conclude that

$$f_x(t,x^*,\dot{x}^*) + \lambda g_x(t,x^*,\dot{x}^*) = \frac{d}{dt}(f_{\dot{x}}(t,x^*,\dot{x}^*) + \lambda g_{\dot{x}}(t,x^*,\dot{x}^*)).$$

(Note that we get the same conclusion from either of the two equations.) This is just the usual Euler–Lagrange equation applied to the *Lagrangian integrand*:

$$L(t,x,\dot{x}) = f(t,x,\dot{x}) + \lambda g(t,x,\dot{x})$$

in terms of the Lagrange multiplier $\lambda$, which leads us to the following solution procedure.
1. Construct the Lagrangian integrand

$$L(t,x,\dot{x}) = f(t,x,\dot{x}) + \lambda g(t,x,\dot{x}).$$

2. Simplify (and ultimately solve) at least one of the following differential equations to produce a general extremal $x^*$:
    - If $L_x = 0$, then you can use the (simplified) Euler–Lagrange equation

    $$L_{\dot{x}} = \text{constant}.$$

    - If $L_t = 0$, then you can use the (simplified) integral Euler–Lagrange equation

    $$L - \dot{x}L_{\dot{x}} = \text{constant}.$$

    - You can always use the Euler–Lagrange equation

    $$L_x = \frac{d}{dt}(L_{\dot{x}}).$$

3. Use any endpoint conditions in the original problem statement to reduce the number of arbitrary constants in your general expression for $x^*$.
4. If there is a hits-the-target condition $x^*(T) = \text{tar}(T)$, then simplify that equation as much as possible and also simplify the Lagrangian transversality condition

$$L(T^*) + (\text{tar}'(T^*) - \dot{x}^*(T^*))L_{\dot{x}}(T^*) = 0.$$

5. Use the integral constraint (and the two equations from the preceding step if they apply) to determine any remaining constants (in particular, the multiplier $\lambda$).

**Lab Activity 3.** *Use Lagrange multipliers to find the extremal satisfying the endpoint conditions and the integral constraint for Dido's isoperimetric problem*

$$\text{Minimize} \int_0^T -x \, dt \quad \text{over } x \in C^2 \text{ on } [0, T]$$

$$\text{with } x(0) = 0, x(T) = 0,$$

$$\text{and } \int_0^T \sqrt{1 + \dot{x}^2} \, dt = 100.$$

*Graph the extremal and find the area of Carthage. (Hint: solve the special case of the integral Euler–Lagrange equation $L - \dot{x}L_{\dot{x}} = 0$, which results from transversality in this situation since $\tan'(T^*) = 0$.)*

## Exercises

7-1. Find the extremal satisfying the endpoint conditions and the integral constraint for

$$\text{Minimize} \int_0^2 \dot{x}^2 \, dt \quad \text{over } x \in C^2 \text{ on } [0, 2]$$

$$\text{with } x(0) = 0, x(2) = 1,$$

$$\text{and } \int_0^2 x \, dt = 1.$$

7-2. Find the extremal satisfying the endpoint conditions and the integral constraint for

$$\text{Minimize} \int_0^\pi \dot{x}^2 \, dt \quad \text{over } x \in C^2 \text{ on } [0, \pi]$$

$$\text{with } x(0) = 0, x(\pi) = 0,$$

$$\text{and } \int_0^\pi x \sin(t) \, dt = \frac{\pi}{4}.$$

7-3. Find the infinite family of extremals satisfying the endpoint conditions and the integral constraint for

$$\text{Minimize} \int_0^\pi \dot{x}^2 \, dt \quad \text{over } x \in C^2 \text{ on } [0, \pi]$$

$$\text{with } x(0) = 0, x(\pi) = 0,$$

$$\text{and } \int_{0}^{\pi} -x^2 \, dt = -\pi.$$

(Hint: do not enforce the second endpoint condition at first and then enforce it by inspection.)

7-4. Use calculus of variations techniques to approximate the extremal satisfying the endpoint conditions and the integral constraint for the reverse Dido problem

$$\text{Minimize } \int_{0}^{T} \sqrt{1 + \dot{x}^2} \, dt \quad \text{over } x \in C^2 \text{ on } [0, T]$$

$$\text{with } x(0) = 0, \, x(T) = 0,$$

$$\text{and } \int_{0}^{T} x \, dt = 1591.$$

(Hint: solve the special case of the integral Euler–Lagrange equation $L - \dot{x}L_{\dot{x}} = 0$ and zoom in on the graph of the constraint integral as a function of $\lambda$ to approximate the multiplier.)

Part II: **Optimal control**

# 8 Discrete-time optimal control

Discrete-time optimal control problems arise in systems that evolve periodically, where the relationships between consecutive periods are described by "difference equations" (discrete surrogates for differential equations). We illustrate this via the "life history" of an organism as dictated by natural selection (adapted from [5]).

## 8.1 Life history

The life history problem for a species refers to the reproduction and mortality schedule favored by natural selection. This schedule is determined by controlling reproductive effort in each of a sequence of discrete age classes in such a way as to maximize the reproductive value of the species, which involves trade-offs between growth, reproduction, and survival.

### 8.1.1 Age classes for elephants

At the end of any year $t$, an elephant population can be grouped into discrete age classes by $x_i(t)$ = number of elephants aged $i$ years (for $i = 0, 1, \ldots, 79$). Figure 8.1 shows individuals who appear to be from different age classes. Individual elephants move through these age classes in such a way that the number in each class for the following year $t + 1$ can be modeled via the following difference equation using a fraction $a_{i+1,i} \in [0, 1]$ of the preceding class from the preceding year (with one exception):

$$x_{i+1}(t + 1) = a_{i+1,i} x_i(t) \quad \text{for } i = 0, 2, \ldots, 78. \tag{8.1}$$

**Question 8.1.** *What is the difference equation for the exception $x_0(t + 1)$?*

**Another grouping**
Of course, we could also group elephants into other age classes. For instance, if we use decade-long age classes

$$x_0(t) = \text{number of elephants aged 0–9,}$$
$$x_1(t) = \text{number of elephants aged 10–19,}$$

---

1 Image Credit: Weldon Kennedy (https://commons.wikimedia.org/wiki/Category:3_elephants#/media/File:202008-_Porini_Mara_Trip-10_(50232570036).jpg) CC BY 2.0 (https://creativecommons.org/licenses/by/2.0/).

https://doi.org/10.1515/9783111290157-008

**Figure 8.1:** Elephants from different age classes.[1]

$$\vdots$$

$$x_7(t) = \text{number of elephants aged 70–79,}$$

then the relationships between age classes generate new difference equations

$$x_0(t + 1) = a_{0,0}x_0(t) + a_{0,1}x_1(t) + a_{0,2}x_2(t) + a_{0,3}x_3(t) + a_{0,4}x_4(t),$$
$$x_1(t + 1) = a_{1,0}x_0(t) + a_{1,1}x_1(t),$$

$$\vdots$$

$$x_7(t + 1) = a_{7,6}x_6(t) + a_{7,7}x_7(t),$$

where the first age class $x_0(t + 1)$ inherits its members from only the first five age classes of the preceding year (assuming that elephants do not reproduce after the age of 49).

**Question 8.2.** *Explain the relationships behind the difference equation $x_1(t + 1) = a_{1,0}x_0(t) + a_{1,1}x_1(t)$.*

**The least-refined grouping**
Perhaps, the least-refined grouping of interest is

$$x_0(t) = \text{\# of elephants who are not yet producing calves,}$$
$$x_1(t) = \text{\# of elephants who are producing calves,} \tag{8.2}$$
$$x_2(t) = \text{\# of elephants who are no longer producing calves.}$$

**Problem 8.1.** *Write the three difference equations for this grouping.*

### 8.1.2 State equation

For any grouping of individuals into discrete age classes

$$x_0(t), x_1(t), \ldots, x_T(t),$$

we can use the vector notation

$$\mathbf{x}(t) = \begin{bmatrix} x_0(t) \\ x_1(t) \\ \vdots \\ x_T(t) \end{bmatrix} \quad \text{and} \quad \mathbf{x}(t+1) = \begin{bmatrix} x_0(t+1) \\ x_1(t+1) \\ \vdots \\ x_T(t+1) \end{bmatrix}$$

to rewrite the corresponding difference equations as a single matrix "state equation"

$$\mathbf{x}(t+1) = A \cdot \mathbf{x}(t) \tag{8.3}$$

in terms of the $(T{+}1){\times}(T{+}1)$ matrix $A$ with entries $a_{i,j}$. The *state equation* in any discrete-time optimal control problem is a matrix difference equation like this that governs the relationship between the age classes (i. e., the *states* of the system) from one year to the next. For instance, the state equation for our least-refined grouping of elephants is

$$\mathbf{x}(t+1) = \begin{bmatrix} a_{0,0} & a_{0,1} & 0 \\ a_{1,0} & a_{1,1} & 0 \\ 0 & a_{2,1} & a_{2,2} \end{bmatrix} \cdot \mathbf{x}(t).$$

In the long-term (where natural selection does its "optimal" work), the state equation (8.3) amounts to a multiplication of the state $\mathbf{x}(t)$ by a number $\lambda$ (the largest eigenvalue of the matrix $A$) to determine the next year state $\mathbf{x}(t+1)$.

**Question 8.3.** *What would you expect about the size of the "ultimate population multiplier" $\lambda$ in the long-term vector equation $\mathbf{x}(t+1) = \lambda\mathbf{x}(t)$ if the species is* not *going extinct?*

Natural selection essentially maximizes the value of the ultimate population multiplier $\lambda$.

### 8.1.3 Controls and cost function

In a general discrete-time optimal control problem, there is a cost function being minimized over all possible *controls*, which are the elements that the agent in the problem can choose. In the case of the life history problem, there is a distinct control $u_i$ for each age class,

$$u_i = \text{the reproductive effort of the individuals in the } i\text{th age class,}$$

satisfying $0 \le u_i \le 1$. Natural selection thus chooses the full vector $\mathbf{u} = (u_1, u_2, \ldots, u_T)$ of these efforts to maximize the ultimate population multiplier $\lambda$ resulting from that choice.

The "cost" in this case is actually a reward to be maximized (which we can convert to a cost to be minimized by taking its negative).

We can reformulate this optimization problem by considering the following two important measures:

- *fecundity*: $\text{fec}_i(\mathbf{u})$ = the number of calves, from an individual in the $i$th age class, that survive long enough to breed, and
- *survivorship*: $\text{sur}_i(\mathbf{u})$ = probability that an individual survives into the $i$th age class.

**Question 8.4.** *What does the product $\text{sur}_i(\mathbf{u}) \, \text{fec}_i(\mathbf{u})$ measure?*

It turns out that the natural selection problem of maximizing the ultimate population multiplier $\lambda$ is equivalent to maximizing the *reproductive value* of the species:

$$J(\mathbf{u}) = \sum_{i=0}^{T} \frac{\text{sur}_i(\mathbf{u}) \, \text{fec}_i(\mathbf{u})}{\lambda^{i+1}}, \tag{8.4}$$

where each term in the sum is discounted by an appropriate power of $\lambda$, since this is essentially a present value (this discounting makes the most sense when consecutive age classes are separated by the same unit of time as the states of the system).

Maximizing the reproductive value $J(\mathbf{u})$ of the species is a nontrivial problem, since a greater reproductive effort $u_i$ in the $i$th age class generally leads to lower values of both fecundity and survivorship in later age classes. Notice however that reproductive effort in any particular age class $\bar{i}$ has no bearing on the fecundity or survivorship associated with earlier age classes $i < \bar{i}$, since reproductive effort in the present only effects fecundity and survivorship in the future. This fact is a key to the solution technique for the life history problem, which we will investigate now.

### 8.1.4 Solving the life history problem

One way to solve the life history problem uses the following definition of the reproductive value of the species for the $i$th age class:

$$J_i(\mathbf{u}) = \frac{\lambda^i}{\text{sur}_i(\mathbf{u})} \sum_{k=i}^{T} \frac{\text{sur}_k(\mathbf{u}) \, \text{fec}_k(\mathbf{u})}{\lambda^{k+1}}. \tag{8.5}$$

Notice that the reproductive value $J$ we already defined in (8.4) satisfies $J = J_0$, since if all individuals survive into the 0th age class, then $\text{sur}_0(\mathbf{u}) = 1$. To solve for the maximum reproductive value $J(\mathbf{u})$, we utilize the following general recursion relationship:

$$J_i(\mathbf{u}) = \frac{1}{\lambda}\left( \text{fec}_i(\mathbf{u}) + \frac{\text{sur}_{i+1}(\mathbf{u})}{\text{sur}_i(\mathbf{u})} J_{i+1}(\mathbf{u}) \right), \tag{8.6}$$

which follows from definition (8.5), as we will see in the next subsection.

Since the values $(u_{i+1}, u_{i+2}, \ldots, u_T)$ of reproductive effort in age classes beyond the $i$th one have no bearing on the fecundity or survivorship associated with earlier age classes, we can first choose $u_T$ to maximize $J_T$ via formula (8.5) and then use (8.6) recursively (backward) to choose $u_{T-1}$ to maximize $J_{T-1}$, then choose $u_{T-2}$ to maximize $J_{T-2}$, and so on until we have chosen every element in the vector $\mathbf{u}$ of controls to maximize $J_0 = J$. This kind of backward recursion technique is called "dynamic programming" [1].

**Example 5.** To illustrate the dynamic programming method, we will maximize the reproductive value of the least-refined grouping of elephants (8.3), given formulas for fecundity and survivorship as in Table 8.1.

**Table 8.1:** Fecundity and survivorship for least-refined grouping.

| $i$ | $fec_i(u_0, u_1, u_2)$ | $sur_i(u_0, u_1, u_2)$ |
|---|---|---|
| 0 | 0 | 1 |
| 1 | $7u_1$ | 0.9 |
| 2 | 0 | $0.8 - 0.2u_1$ |

Starting from the last age class $i = 2$, we know from (8.5) that

$$J_2(\mathbf{u}) = \frac{\lambda^2}{sur_2(\mathbf{u})} \frac{sur_2(\mathbf{u}) \, fec_2(\mathbf{u})}{\lambda^3} = \frac{fec_2(\mathbf{u})}{\lambda} = 0,$$

since $fec_2(\mathbf{u}) = 0$. Thus $J_2$ is maximized (with maximum value 0) at any $u_2$, which means that elephants in the last group can choose any value of reproductive effort $0 \le u_2 \le 1$ they like without affecting the reproductive value of their species. This makes sense since the reproductive effort of individuals who cannot produce calves in the future is irrelevant to the reproductive value of the species.

It follows from (8.6) that

$$J_1(\mathbf{u}) = \frac{1}{\lambda}\left( fec_1(\mathbf{u}) + \frac{sur_2(\mathbf{u})}{sur_1(\mathbf{u})} J_2(\mathbf{u}) \right) = \frac{fec_1(\mathbf{u})}{\lambda} = \frac{7u_1}{\lambda},$$

which is maximized over $0 \le u_1 \le 1$ by choosing $u_1 = 1$. This results in a maximum value of $J_1(u_0, 1, u_2) = \frac{7}{\lambda}$.

Finally, it follows from (8.6) that

$$J_0(\mathbf{u}) = \frac{1}{\lambda}\left( fec_0(\mathbf{u}) + \frac{sur_1(\mathbf{u})}{sur_0(\mathbf{u})} J_1(\mathbf{u}) \right)$$

$$\Downarrow$$

$$J_0(u_0, 1, u_2) = \frac{1}{\lambda}\left( 0 + \frac{0.9}{1} \frac{7}{\lambda} \right) = \frac{6.3}{\lambda^2},$$

where we substitute $u_1 = 1$, since it is the maximizing value of $u_1$ we determined in the preceding step. The ratio $\frac{6.3}{\lambda^2}$ is maximized at any value of $u_0$ (since it does not depend on $u_0$), so we know that $J_0$ is maximized at $(u_0, 1, u_2)$ for any choices of $u_0$ and $u_2$ satisfying $0 \le u_0 \le 1$ and $0 \le u_2 \le 1$ (with maximum value $\frac{6.3}{\lambda^2}$). We conclude that the reproductive value of this grouping is maximized if the middle group gives maximum reproductive effort, regardless of the effort given by the other two groups. This makes sense since the middle group is the only one producing calves, and the reproductive effort in the two other groups has no effect on fecundity or survivorship of the first or middle group.

**Problem 8.2.** *Carry out dynamic programming to maximize the reproductive value of the following grouping of elephants:*

$$x_0(t) = \text{\# of elephants aged 0--11,}$$
$$x_1(t) = \text{\# of elephants aged 12--30,}$$
$$x_2(t) = \text{\# of elephants aged 31--45,}$$
$$x_3(t) = \text{\# of elephants aged 46--79,}$$

*given formulas for fecundity and survivorship as in Table 8.2 and assuming that $\lambda \ge 1$.*

**Table 8.2:** Fecundity and survivorship for Problem 8.2.

| $i$ | $\mathrm{fec}_i(u_0, u_1, u_2, u_3)$ | $\mathrm{sur}_i(u_0, u_1, u_2, u_3)$ |
|---|---|---|
| 0 | 0 | 1 |
| 1 | $7u_1$ | 0.9 |
| 2 | $7u_2 - 2u_1$ | $0.8 - 0.2u_1$ |
| 3 | 0 | $0.7 - 0.3u_2 - 0.1u_1$ |

### 8.1.5 Proving the general recursion relationship (8.6)

From formula (8.5) applied with $i + 1$ in place of $i$ we have

$$J_{i+1}(\mathbf{u}) = \frac{\lambda^{i+1}}{\mathrm{sur}_{i+1}(\mathbf{u})} \sum_{k=i+1}^{T} \frac{\mathrm{sur}_k(\mathbf{u})\,\mathrm{fec}_k(\mathbf{u})}{\lambda^{k+1}}.$$

Multiplying this by $\frac{\mathrm{sur}_{i+1}(\mathbf{u})}{\mathrm{sur}_i(\mathbf{u})}$ gives

$$\frac{\mathrm{sur}_{i+1}(\mathbf{u})}{\mathrm{sur}_i(\mathbf{u})} J_{i+1}(\mathbf{u}) = \frac{\lambda^{i+1}}{\mathrm{sur}_i(\mathbf{u})} \sum_{k=i+1}^{T} \frac{\mathrm{sur}_k(\mathbf{u})\,\mathrm{fec}_k(\mathbf{u})}{\lambda^{k+1}}.$$

Adding $\mathrm{fec}_i(\mathbf{u})$ to this gives

$$\text{fec}_i(\mathbf{u}) + \frac{\text{sur}_{i+1}(\mathbf{u})}{\text{sur}_i(\mathbf{u})} J_{i+1}(\mathbf{u}) = \text{fec}_i(\mathbf{u}) + \frac{\lambda^{i+1}}{\text{sur}_i(\mathbf{u})} \sum_{k=i+1}^{T} \frac{\text{sur}_k(\mathbf{u})\,\text{fec}_k(\mathbf{u})}{\lambda^{k+1}}.$$

Multiplying and dividing the first term on the right by $\lambda^{i+1}\,\text{sur}_i(\mathbf{u})$ gives

$$\text{fec}_i(\mathbf{u}) + \frac{\text{sur}_{i+1}(\mathbf{u})}{\text{sur}_i(\mathbf{u})} J_{i+1}(\mathbf{u}) = \frac{\lambda^{i+1}\,\text{sur}_i(\mathbf{u})\,\text{fec}_i(\mathbf{u})}{\lambda^{i+1}\,\text{sur}_i(\mathbf{u})} + \frac{\lambda^{i+1}}{\text{sur}_i(\mathbf{u})} \sum_{k=i+1}^{T} \frac{\text{sur}_k(\mathbf{u})\,\text{fec}_k(\mathbf{u})}{\lambda^{k+1}}.$$

Factoring out $\frac{\lambda^{i+1}}{\text{sur}_i(\mathbf{u})}$ on the right side gives

$$\text{fec}_i(\mathbf{u}) + \frac{\text{sur}_{i+1}(\mathbf{u})}{\text{sur}_i(\mathbf{u})} J_{i+1}(\mathbf{u}) = \frac{\lambda^{i+1}}{\text{sur}_i(\mathbf{u})} \left( \frac{\text{sur}_i(\mathbf{u})\,\text{fec}_i(\mathbf{u})}{\lambda^{i+1}} + \sum_{k=i+1}^{T} \frac{\text{sur}_k(\mathbf{u})\,\text{fec}_k(\mathbf{u})}{\lambda^{k+1}} \right),$$

which simplifies to

$$\text{fec}_i(\mathbf{u}) + \frac{\text{sur}_{i+1}(\mathbf{u})}{\text{sur}_i(\mathbf{u})} J_{i+1}(\mathbf{u}) = \frac{\lambda^{i+1}}{\text{sur}_i(\mathbf{u})} \sum_{k=i}^{T} \frac{\text{sur}_k(\mathbf{u})\,\text{fec}_k(\mathbf{u})}{\lambda^{k+1}} = \lambda J_i(\mathbf{u}).$$

Dividing both sides of the resulting equation gives (8.6), as claimed.

## Exercises

8-1. Carry out dynamic programming to maximize the reproductive value of the same grouping of elephants as in Problem 8.2 given the same formulas for fecundity and survivorship Table 8.2, but assuming that $\lambda < 1$.
(Hint: you should identify two different maximizing strategies, depending on whether $\lambda$ is above or below the threshold value $\bar{\lambda} = \sqrt[3]{\frac{26}{63}}$.)

8-2. Carry out dynamic programming to maximize the reproductive value of the following grouping of elephants:

$$x_0(t) = \text{\# of elephants aged 0–11,}$$
$$x_1(t) = \text{\# of elephants aged 12–20,}$$
$$x_2(t) = \text{\# of elephants 21–31,}$$
$$x_3(t) = \text{\# of elephants aged 32–45,}$$
$$x_4(t) = \text{\# of elephants aged 46–79,}$$

given formulas for fecundity and survivorship as in Table 8.3 and assuming that $\lambda \geq 1$.

**Table 8.3:** Fecundity and survivorship for Exercise 8-2.

| $i$ | $\text{fec}_i(u_0, u_1, u_2, u_3, u_4)$ | $\text{sur}_i(u_0, u_1, u_2, u_3, u_4)$ |
|---|---|---|
| 0 | 0 | 1 |
| 1 | $6u_1$ | 0.9 |
| 2 | $8u_2 - 2u_1$ | $0.8 - 0.2u_1$ |
| 3 | $5u_3 - 2u_2 - 4u_1$ | $0.7 - 0.3u_2 - 0.1u_1$ |
| 4 | 0 | $0.6 - 0.4u_3 - 0.2u_2 - 0.1u_1$ |

# 9 Continuous-time optimal control

In the preceding chapter, we saw how a discrete-time optimal control problem could be used to model the effect of natural selection on the life history of an organism, as long as that organism was categorized into discrete age classes. This model turns out to be not so accurate for organisms (like plants) whose responses to their environment are not constant throughout the time span of an age class, but instead vary continuously [5]. However, the model can be extended to cover this situation as well.

## 9.1 Life history of plants

There are at least three time-varying controls for an annual plant species [6]:

$$u_1(t) = \text{fraction of fixed carbon allocated to leaves,}$$
$$u_2(t) = \text{fraction of fixed carbon stored,}$$
$$u_3(t) = \text{fraction of fixed carbon allocated to seeds,}$$

which affect the state $\mathbf{x}(t)$ of a representative plant at each time $t$:

$$x_1(t) = \text{number of leaves,}$$
$$x_2(t) = \text{amount of stored energy,}$$
$$x_3(t) = \text{number of seeds}$$

via three (differential) *state equations*

$$\dot{x}_1 = \frac{r}{c_1} x_1 u_1,$$
$$\dot{x}_2 = \frac{r}{c_2} x_1 u_2,$$
$$\dot{x}_3 = \frac{r}{c_3} x_1 u_3$$

in terms of the parameters

$$r = \text{rate of photosynthesis,}$$
$$c_1 = \text{cost of a new leaf,}$$
$$c_2 = \text{cost of storing energy,}$$
$$c_3 = \text{cost of a seed.}$$

The measure of reproductive value for an annual plant is simply the total number of seeds produced during its life cycle, and this is the quantity that natural selection seeks to maximize.

https://doi.org/10.1515/9783111290157-009

**Life history of perennials**

The model for annual plants can be adapted to perennials by considering each year in the life of the plant as a different age class and using age-specific control functions $\mathbf{u}_i(t) = (u_{i,1}(t), u_{i,2}(t), u_{i,3}(t))$ and corresponding states $\mathbf{x}_i(t) = (x_{i,1}(t), x_{i,2}(t), x_{i,3}(t))$ for each age class $i$. This leads to a discrete structure overarching the continuous model within each age class, so we can use many of the techniques for discrete-time optimal control problems.

To begin our study of general continuous-time optimal control problems, we will instead focus on a simpler model in the following section.

## 9.2 Single-state continuous-time optimal control

Consider a single state function $x(t)$ and a single control function $u(t)$ whose outputs belong to some bounded set $U$. In this case the general continuous-time optimal control problem is to minimize the *cost functional*

$$J[u] = \int_{t_0}^{T} f_0(x, u) \, dt$$

over control functions $u$ satisfying $u(t) \in U$, where the end time $T$ may be free, and where the state $x$ is determined by the *state equation* $\dot{x} = f(x, u)$ and the endpoint conditions $x(t_0) = x_0$ and $x(T) = x_1$.

For example, natural selection attempts to maximize the number of seeds $x_3(T)$ at the end $T$ of the life cycle of an annual plant. We can write this as a minimization of $-x_3(T)$, and (using $x_3(0) = 0$) we can use the fundamental theorem of calculus to rewrite this as an integral

$$-x_3(T) = \int_{0}^{T} -\dot{x}_3 \, dt = \int_{0}^{T} \underbrace{-\frac{r}{c_3} x_1 u_3}_{f_0(x,u)} \, dt,$$

where we have applied the state equation $\dot{x}_3 = \frac{r}{c_3} x_1 u_3$. However, there are multiple state equations in the life history problem for plants, so its cost functional is really the only part of that problem fitting our simpler single-state model. Now we will explore a vaccination problem whose components all fit the simpler model.

**Problem 9.1** (Vaccination problem). *Suppose $x(t)$ is the number of people infected by a virus at time $t$ and that, without intervention, the infection would spread at a rate proportional to how many people are infected. To combat the infection, we can vaccinate at a rate of $u(t)$, which cannot exceed some fixed upper bound $M > 0$. Identify the components $U$, $f_0(x, u)$, and $f(x, u)$ of the optimal control problem whose solution is the vaccination rate that moves the number of infected people from $x_0$ to $x_1$ as quickly as possible.*

### 9.2.1 The state equation as a constraint

**Question 9.1.** *How many equality constraints*

$$\dot{x}(t) - f\big(x(t), u(t)\big) = 0$$

*does the state equation effectively represent?*

Recall our approach from constrained calculus of variations problems, where we had a single multiplier for a single integral constraint. To fit that context, we could create a single integral constraint from the state equation simply by integrating it:

$$\int_{t_0}^{T} \dot{x}(t) - f\big(x(t), u(t)\big) \, dt = 0.$$

However, we need a separate multiplier $\lambda(t)$ for each integrand $\dot{x}(t) - f(x(t), u(t))$ in this case, since these integrands need to be zero for every $t$ (and not just have their integral equal zero). Therefore we instead include a multiplier function $\lambda(t)$ before incorporating the state equation directly into the Lagrangian integrand:

$$\int_{t_0}^{T} f_0(x, u) + \lambda \cdot (\dot{x} - f(x, u)) \, dt = \int_{t_0}^{T} \lambda \cdot \dot{x} + \underbrace{(f_0(x, u) - \lambda \cdot f(x, u))}_{H(\lambda, x, u)} \, dt,$$

which we have rewritten in terms of the *Hamiltonian* function $H$ (named after the Irish mathematician Sir William Rowan Hamilton). Notice that $H$ identifies $\lambda$ explicitly as one of its arguments, and remember that $x$, $u$, and $\lambda$ are all functions of $t$.

**Problem 9.2.** *By considering $\lambda$ and $u$ to be independent of $x$ and $\dot{x}$, what does the Euler–Lagrange equation applied to the integrand $\lambda \cdot \dot{x} + H(\lambda, x, u)$ yield in terms of $H$?*

*What does the transversality condition applied to the integrand $\lambda \cdot \dot{x} + H(\lambda, x, u)$ yield in terms of $H$ (assuming that $T$ is free)?*

The equation that results from the Euler–Lagrange equation in this case is called the *costate equation*, as it determines the *costate* function $\lambda(t)$, which is a kind of partner to the state function $x(t)$.

### 9.2.2 Pontryagin's principle

The key principle for choosing an optimal control is credited to the Russian mathematician Lev Semenovich Pontryagin, and this principle can be illustrated by finding the control function $u(t)$ satisfying $u(t) \in [-1, 1]$ that minimizes the integral $\int_0^4 p(t)u(t) \, dt$ for some continuous function $p(t)$. An optimal control in this case is

$$u^*(t) = \begin{cases} -1 & \text{if } p(t) > 0, \\ 1 & \text{if } p(t) \le 0, \end{cases}$$

which should be clear from the example shown in Figure 9.1. The idea is that for each fixed $t$, we choose $u^*(t)$ to minimize the integrand evaluated at that fixed $t$. Notice that the control function is not necessarily continuous.

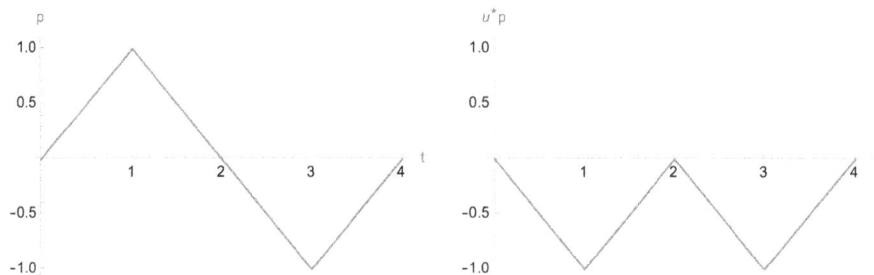

**Figure 9.1:** Graphs of $p(t)$ and $u^*(t)p(t)$.

Pontryagin applied this to the Lagrangian integral

$$\int_{t_0}^{T} \lambda(t) \cdot \dot{x}(t) + H(\lambda(t), x(t), u(t)) \, dt$$

and focused entirely on the Hamiltonian since that is the only part of the Lagrangian integrand that depends on the control $u$:

**Pontryagin's Principle**

> Choose $u^*$ so that for each $t \in [t_0, T]$, $u^*(t)$ minimizes $H(\lambda(t), x(t), u)$ with respect to $u$.

### 9.2.3 Solution procedure

The following solution procedure summarizes what we have learned in this chapter:

1. Build the Hamiltonian function

$$H(\lambda, x, u) = f_0(x, u) - \lambda f(x, u).$$

2. Use the costate equation $\dot{\lambda} = H_x$, Pontryagin's principle, and the state equation $\dot{x} = f(x, u)$ to find the optimal costate $\lambda^*(t)$, control $u^*(t)$, and state $x^*(t)$.

3. Use endpoint conditions and transversality $H(T^*) = 0$ (when $T$ is free) to determine any arbitrary constants.

**Problem 9.3.** *Use the solution procedure to solve the vaccination problem. (Note that your solution should come with cases depending on relationships between $x_0$ and $x_1$.)*

### 9.2.4 Solving coupled differential equations

Some of the problems you encounter in optimal control will have "coupled" state and costate equations whose (unknown) functions $x(t)$ and $\lambda(t)$ may appear in both equations. Here is an example of one such coupled pair of differential equations:

$$\dot{x} = x + \lambda,$$
$$\dot{\lambda} = x - \lambda,$$

which we can solve in Python or Mathematica via slight modifications of the commands we used in Section 2.3.

**Python**

```
#import functions and define variables
from sympy import symbols, Function, dsolve, diff, Eq
t = symbols("t")
x = Function('x')
lam = Function('lam')

#solve the coupled equations
print(dsolve([Eq(diff(x(t),t),x(t)+lam(t)),Eq(diff(lam(t),t)
    ,x(t)-lam(t))]))
```

Note in particular the square brackets [ ] containing both differential equations inside the dsolve command.

**Mathematica**

```
DSolve[{x'[t] == lam[t] + x[t], lam'[t] == x[t] - lam[t]}, {
    x[t], lam[t]}, t]
```

Note in particular the curly braces { } containing both differential equations inside the Dsolve command.

## Exercises

9-1. Give the complete optimal control problem that results if we modify the goal in the vaccination problem to be minimizing the total amount of vaccination from time 0 to time $T$.

9-2. Assuming that $x_1 < x_0 < M$, use the solution procedure to find the optimal control $u^*$ and end time $T^*$ that minimize $\int_0^T u\,dt$ over controls $u$ satisfying $u(t) \in [0,M]$ for $t \in [0,T]$ with state equation $\dot{x} = x - u$ and endpoint conditions $x(0) = x_0$ and $x(T) = x_1$.

9-3. In this exercise, we consider modified versions of the vaccination problem where the vaccination rate $u$ is bounded instead by $-M \le u \le M$ (for $M > 1$). Use the solution procedure to find the optimal control $u^*$ and end time $T^*$ (if they exist) when the endpoint conditions are $x(0) = 1$ and $x(T) = 0$, and the cost to be minimized is

 (i) $\int_0^T |u|\,dt$.

 (ii) $\int_0^T \frac{1}{4}u^4\,dt$.

9-4. Use the solution procedure to find the optimal control $u^*$ and end time $T^*$ for the modified version of the vaccination problem where the vaccination rate $u$ is not subject to bounds, the initial state $x_0$ and target state $x_1$ satisfy $x_0 > x_1$, and the cost to be minimized is

$$\int_0^T \frac{x^2}{2} + \frac{u^2}{2}\,dt.$$

# 10 Multistate optimal control

In the preceding chapter, we developed a procedure for solving optimal control problems with one state and one control. Now we will extend that procedure to optimal control problems with multiple states $\mathbf{x}(t)$ (each component $x_i(t)$ of the vector $\mathbf{x}(t)$ represents a different state) and (potentially) multiple controls $\mathbf{u}$:

$$\text{Minimize } J[\mathbf{u}] = \int_{t_0}^{T} f_0(\mathbf{x}, \mathbf{u})\, dt \quad \text{over controls } \mathbf{u} \text{ with } \mathbf{u}(t) \in U,$$

where the end time $T$ may be free, and where each state $x_i$ is determined by a corresponding state equation $\dot{x}_i(t) = f_i(\mathbf{x}(t), \mathbf{u}(t))$ and the endpoint conditions $\mathbf{x}(t_0) = \mathbf{x}_0$ and $\mathbf{x}(T) = \mathbf{x}_1$.

## 10.1 Multistate solution procedure

To extend our results to multiple states $\mathbf{x}(t)$, we use vector-function costates $\boldsymbol{\lambda}(t)$ (of the same dimension as $\mathbf{x}(t)$) defined by the costate equations

$$\dot{\lambda}_i = \frac{\partial H}{\partial x_i}.$$

If there are also multiple controls $\mathbf{u}(t)$, then we apply Pontryagin's principle to the corresponding multivariable Hamiltonian. Otherwise, the following solution procedure is very similar to that for single-state optimal control:

1. Build the Hamiltonian function

$$H(\boldsymbol{\lambda}, \mathbf{x}, \mathbf{u}) = f_0(\mathbf{x}, \mathbf{u}) - \lambda_1 f_1(\mathbf{x}, \mathbf{u}) - \lambda_2 f_2(\mathbf{x}, \mathbf{u}) - \cdots - \lambda_m f_m(\mathbf{x}, \mathbf{u}),$$

   where $m$ is the number of states.
2. Use the $m$ costate equations $\dot{\lambda}_i = \frac{\partial H}{\partial x_i}$, Pontryagin's principle, and the $m$ state equations $\dot{x}_i = f_i(\mathbf{x}, \mathbf{u})$ to find the optimal $\boldsymbol{\lambda}^*(t)$, $\mathbf{u}^*(t)$, and $\mathbf{x}^*(t)$.
3. Use endpoint conditions and transversality $H(T^*) = 0$ (when $T$ is free) to determine any arbitrary constants.

**Example 6.** In this example, we want to find the time-optimal control $u$ satisfying $|u(t)| \leq 1$ that moves the states $(x_1, x_2)$ from $(x_1(0), x_2(0)) = (0, 0)$ to $(x_1(T), x_2(T)) = (a, b)$ via the state equations

$$\dot{x}_1 = -x_1 + u,$$
$$\dot{x}_2 = u.$$

https://doi.org/10.1515/9783111290157-010

First, we notice that time is measured by the cost functional $\int_0^T 1\,dt$, which means that our cost integrand is $f_0(x, u) = 1$. Now we are ready to follow the multistate solution procedure outlined above.

1. The Hamiltonian in this case is

$$H(\lambda, \mathbf{x}, \mathbf{u}) = 1 - \lambda_1(-x_1 + u) - \lambda_2 u,$$

where we have used two costates $\lambda_1$ and $\lambda_2$ (one for each component state equation).

2. – The costate equations $\dot{\lambda}_i = \frac{\partial H}{\partial x_i}$ in this case become

$$\dot{\lambda}_1 = \lambda_1 \quad \Longrightarrow \quad \lambda_1^*(t) = c_1 e^t,$$
$$\dot{\lambda}_2 = 0 \quad \Longrightarrow \quad \lambda_2^*(t) = c_2$$

in terms of arbitrary constants $c_1$ and $c_2$.

– Pontryagin's principle leads us to minimizing the Hamiltonian

$$1 - \lambda_1^*(-x_1 + u) - \lambda_2^* u = 1 + c_1 e^t x_1 - \underbrace{(c_1 e^t + c_2)}_{s(t)} u$$

with respect to $u$ (with $|u| \leq 1$). This leads us to the optimal control

$$u^*(t) = \operatorname{sgn}(s(t)) = \begin{cases} 1 & \text{if } s(t) > 0, \\ \in [-1, 1] & \text{if } s(t) = 0, \\ -1 & \text{if } s(t) < 0, \end{cases}$$

where sgn represents the "sign" function, and $s(t) = c_1 e^t + c_2$ is the *switching function* for this problem. Note that in some other contexts, $\operatorname{sgn}(0)$ is defined to be zero, but the control can take any value in $U = [-1, 1]$ in this case, so we alter our definition of sgn accordingly.

The switching function gets its name from the fact that $u^*$ switches from one (extreme) control to the other when $s(t) = 0$ at $t = \ln(\frac{-c_2}{c_1})$. Note that there is no more than one such switch, as long as at least one of $c_1$ or $c_2$ is nonzero, since then $s(t) = c_1 e^t + c_2$ has at most one zero. The situation where $c_1 = c_2 = 0$ can be ruled out since then the Hamiltonian always equals 1 (contradicting $H(T^*) = 0$).

– The state equations $\dot{x}_1 = -x_1 + u^*$ and $\dot{x}_2 = u^*$ then lead to

$$x_1^*(t) = c_3 e^{-t} + \operatorname{sgn}(s(t)),$$
$$x_2^*(t) = \operatorname{sgn}(s(t))t + c_4.$$

3. The initial condition $(x_1(0), x_2(0)) = (0, 0)$ gives us

$$x_1^*(t) = -\operatorname{sgn}(s(0))e^{-t} + \operatorname{sgn}(s(t)),$$
$$x_2^*(t) = \operatorname{sgn}(s(t))t, \tag{10.1}$$

so the other endpoint condition translates into

$$a = -\operatorname{sgn}(s(0))e^{-T^*} + \operatorname{sgn}(s(T^*)), \tag{10.2}$$

$$b = \operatorname{sgn}(s(T^*))T^*. \tag{10.3}$$

Since $T^* > 0$, we conclude from (10.3) that $\operatorname{sgn}(s(T^*)) = \operatorname{sgn}(b)$ and, therefore, that $T^* = |b|$.

According to formula (10.1), the second state $x_2^*$ is a simple linear function with slope either 1 or –1 (and possibly switching between these values). Figure 10.1 shows two possibilities: one using slope 1 the entire time and one switching from slope 1 to –1.

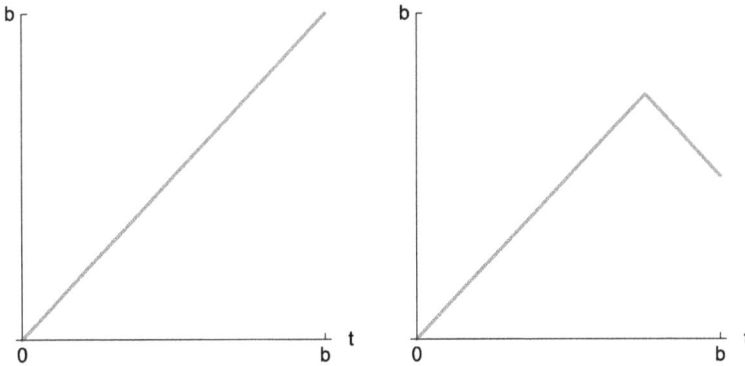

**Figure 10.1:** Two possible graphs for the second state $x_2^*(t)$.

This illustrates the fact that the only way for such a function $x_2^*$ to get from 0 to $b$ on the vertical axis in exactly $T^* = |b|$ time units is if there is a constant slope $\operatorname{sgn}(s(t)) = \operatorname{sgn}(b)$. This means there is no switching in the optimal control $u^*(t) = \operatorname{sgn}(b)$.

The fact that $T^*$ also has to satisfy the first equation in (10.2) then tells us that only target states $(a, b)$ satisfying the equation

$$-\operatorname{sgn}(b)e^{-|b|} + \operatorname{sgn}(b) = a$$

can be reached time-optimally (note that we have used that $\operatorname{sgn}(s(0)) = \operatorname{sgn}(b)$ since we know that there is no switching).

Figure 10.2 shows the curve in the $x_1 x_2$-plane identifying these target states, where the optimal control $u^* = 1$ can be used when the target $(a, b)$ is on the part of the curve in the upper-right quadrant, and the optimal control $u^* = -1$ can be used when the target $(a, b)$ is on the part of the curve in the lower-left quadrant.

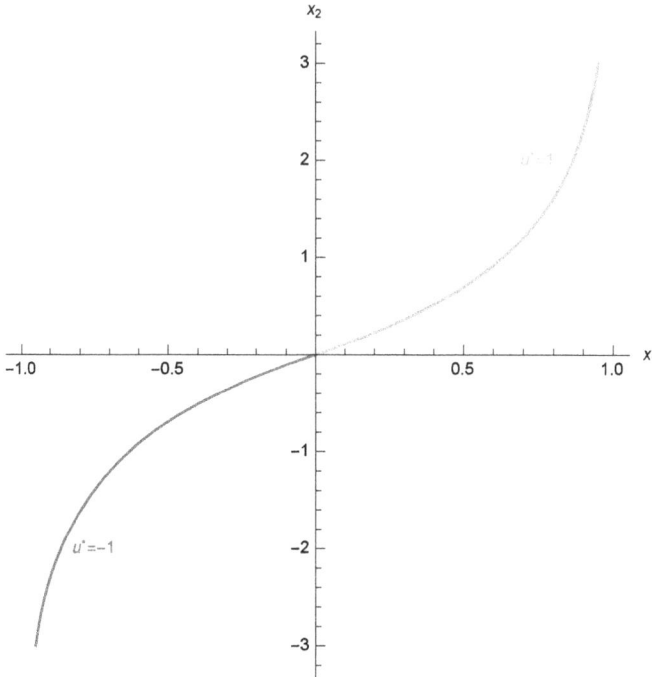

**Figure 10.2:** Time-optimal target states.

## 10.2 End-state costs

Optimal control problems sometimes come with end-state costs, which take the form of some function $g(x)$ applied to the end state $x(T)$ as follows:

$$\text{Minimize } J[u] = \int_{t_0}^{T} f_0(x, u)\, dt + g(x(T)) \quad \text{over } u \text{ with } u(t) \in U,$$

where the end time $T$ is fixed, and the state $x$ is determined by the state equation $\dot{x} = f(x, u)$ and the endpoint condition $x(t_0) = x_0$ (and $x(T)$ is free). For instance, a modified version of the vaccination problem from Problem 9.1 could aim to minimize the number of infected individuals at the end time $T$, in which case the components of the cost would be $f_0(x, u) = 0$ and $g(x) = x$.

For problems with end-state costs, it turns out that our typical analysis is essentially unchanged if we use a "pseudo-costate" $\psi$ in place of a costate $\lambda$ via the formula

$$\psi(t) = \lambda(t) - g'(x(t)). \tag{10.4}$$

This works because we can use the identity

$$\int_{t_0}^{T} g'(x(t))\dot{x}(t)\, dt = g(x(t))\Big|_{t_0}^{T} = g(x(T)) - g(x(t_0)) = g(x(T)) - g(x_0) \quad (10.5)$$

(obtained via the fundamental theorem of calculus applying the endpoint condition $x(t_0) = x_0$) to rewrite the cost as

$$J[u] = \int_{t_0}^{T} f_0(x, u)\, dt + g(x(T))$$

$$= \int_{t_0}^{T} f_0(x, u) + g'(x)\dot{x}\, dt + g(x_0)$$

$$= \int_{t_0}^{T} f_0(x, u) + g'(x)f(x, u)\, dt + g(x_0),$$

where we suppressed the argument $t$ from (10.5) in the second line and applied the state equation $\dot{x} = f(x, u)$ in the last equality. The optimal control $u^*$ is unaffected by the constant $g(x_0)$, so we focus only on the integral portion of the cost:

$$\tilde{J}[u] = \int_{t_0}^{T} f_0(x, u) + g'(x)f(x, u)\, dt. \quad (10.6)$$

Note that the optimal value $J[u^*]$ of the original cost is equal to $\tilde{J}[u^*] + g(x_0)$.

If we build our Hamiltonian for the modified cost functional $\tilde{J}[u]$ in (10.6), then we get

$$H(\lambda, x, u) = f_0(x, u) + g'(x)f(x, u) - \lambda f(x, u) = f_0(x, u) - \psi f(x, u)$$

in terms of the pseudo-costate $\psi$ from (10.4). We then proceed with our analysis in the usual way, generating in particular a "pseudo-costate" equation $\dot{\psi} = \frac{\partial H}{\partial x}$ in place of a costate equation. Note that the transversality condition in this case is applied to the integrand $\lambda \cdot \dot{x} + f_0(x, u) - \psi f(x, u)$ and results in

$$\lambda(T^*) \cdot \dot{x}^*(T^*) + f_0(T^*) - \psi(T^*)f(T^*) + (\text{tar}'(T^*) - \dot{x}^*(T^*)) \cdot \lambda(T^*) = 0$$

$$\Downarrow$$

$$\lambda(T^*) = 0,$$

where we obtained the final implication after dividing by $\text{tar}'(T^*)$, since the target curve is a vertical line through $T^*$ (with infinite slope).

If there are multiple states $\mathbf{x}(t)$ (each component $x_i(t)$ of the vector $\mathbf{x}(t)$ represents a different state) and (potentially) multiple controls $\mathbf{u}$, then the same procedure works if we define multiple pseudo-costates $\psi_i$ as follows:

$$\psi_i(t) = \lambda_i(t) - \frac{\partial g}{\partial x_i}(\mathbf{x}(t)).$$

## Exercises

10-1. Use the multistate solution procedure to solve the time-optimal control problem with state equations

$$\dot{x}_1 = -x_2 + u,$$
$$\dot{x}_2 = x_1 + u,$$

control bound $|u(t)| \le 1$, and endpoint conditions

$$(x_1(0), x_2(0)) = (0, 0),$$
$$(x_1(T), x_2(T)) = (\sqrt{2} - 1, 1).$$

Hint: use the identity

$$a \cos(t) + b \sin(t) = \pm \sqrt{a^2 + b^2} \sin\left(t + \tan^{-1} \frac{a}{b}\right).$$

10-2. Write the fixed-endpoint calculus of variations problem

$$\min \int_{t_0}^{t_1} f(t, x, \dot{x}) \, dt \quad \text{such that } x(t_0) = x_0, x(t_1) = x_1$$

as a two-state optimal control problem using a single control function $u$. Make sure to clearly identify all the necessary components of the optimal control problem and identify the condition from the calculus of variations that results from solving for the optimal control via the costate equations and Pontryagin's principle.

10-3. Use the solution procedure to find the optimal vaccination rate (in a modified version of the vaccination problem from Problem 9.1) that minimizes the number of infected people after $T = 6$ time units.

# 11 Nanosatellite

In this chapter, we consider an example of a nanosatellite (i. e., a small satellite weighing between 1 and 10 kilograms) that naturally leads to a two-state optimal control problem whose optimal controls are dictated by the position of the initial state in the position–velocity plane. As part of this analysis, we will explore an important object in the position–velocity plane called the "switching curve" (no direct relation to the switching function).

## 11.1 Details

Imagine a simple nanosatellite that can move left or right along a horizontal axis via thrusters on its left and right sides. Figure 11.1 shows a nanosatellite moving to the right via a steam thruster on the left. The two thrusters allow us to control the position $x(t)$ of the nanosatellite along the axis via acceleration (left or right) $\ddot{x} = u$ within bounds $|u(t)| \leq 1$.

**Figure 11.1:** Thruster.[1]

**Question 11.1.** *How can we turn the second-order differential equation $\ddot{x} = u$ into a pair of first-order state equations?*

It is convenient to convert our pair of first-order state equations into a single "linear" state equation by using matrix and vector notation as follows:

---

1 Image Credit: NASA (https://www.nasa.gov/sites/default/files/styles/full_width_feature/public/thumbnails/image/ocsd_02_v02.png).

https://doi.org/10.1515/9783111290157-011

$$\begin{bmatrix} \dot{x}_1 \\ \dot{x}_2 \end{bmatrix} = \underbrace{\begin{bmatrix} 0 & 1 \\ 0 & 0 \end{bmatrix}}_{A} \cdot \begin{bmatrix} x_1 \\ x_2 \end{bmatrix} + u \underbrace{\begin{bmatrix} 0 \\ 1 \end{bmatrix}}_{b} = \begin{bmatrix} x_2 \\ u \end{bmatrix}. \tag{11.1}$$

### 11.1.1 Time-optimal nanosatellite

One simple problem involves moving the nanosatellite as quickly as possible from position $a$ with initial velocity $b$ to position 0 at rest (adapted from [2, Example 4.2]). We can encode this as the following optimal control problem:

$$\text{Minimize} \int_{t_0}^{T} 1 \, dt \quad \text{over controls } u \text{ with } u(t) \in [-1, 1],$$

where the state is determined by the state equation (11.1) and the endpoint conditions

$$\begin{bmatrix} x_1(0) \\ x_2(0) \end{bmatrix} = \begin{bmatrix} a \\ b \end{bmatrix} \quad \& \quad \begin{bmatrix} x_1(T) \\ x_2(T) \end{bmatrix} = \begin{bmatrix} 0 \\ 0 \end{bmatrix}.$$

**Problem 11.1.** *Follow the procedure for finding the optimal control and optimal states for this problem, up through the step where you enforce the first endpoint condition (but not the second). Make sure to identify what kind of switching is possible for the optimal control.*

## 11.2 Visualization

For a pair of differential equations governing two states $x_1$ and $x_2$, we can visualize the simultaneous time dependence of the states in the $x_1x_2$-plane, where $x_1$ is on the horizontal axis, and $x_2$ is on the vertical axis, with time as a "hidden" parameter. For the nanosatellite problem, the point $(-10, 6)$ in the $x_1x_2$-plane shown in Figure 11.2 represents the nanosatellite starting at position $-10$ with velocity 6 (moving rightward, since positive). The control used is $u^* = -1$ (leftward acceleration), and the resulting curve shown in Figure 11.2 represents the *state trajectory* pairing the $x_1(t)$ and $x_2(t)$ coordinates as the parameter $t$ changes.

We can see that the values of $x_1$ (the position of the nanosatellite) along this state trajectory increase consistently from $-10$ to 8, at which point $x_1$ decreases consistently from 8 to $-5$. At the same time, the values of $x_2$ (the velocity of the nanosatellite) along this state trajectory decrease consistently from 6 to $-5$ (passing through zero when $x_1 = 8$, and the nanosatellite begins its leftward movement). The control $u^* = -1$ acts as a brake in this case until position $x_1 = 8$ is reached, at which point it acts as an accelerator since the nanosatellite is thereafter moving to the left.

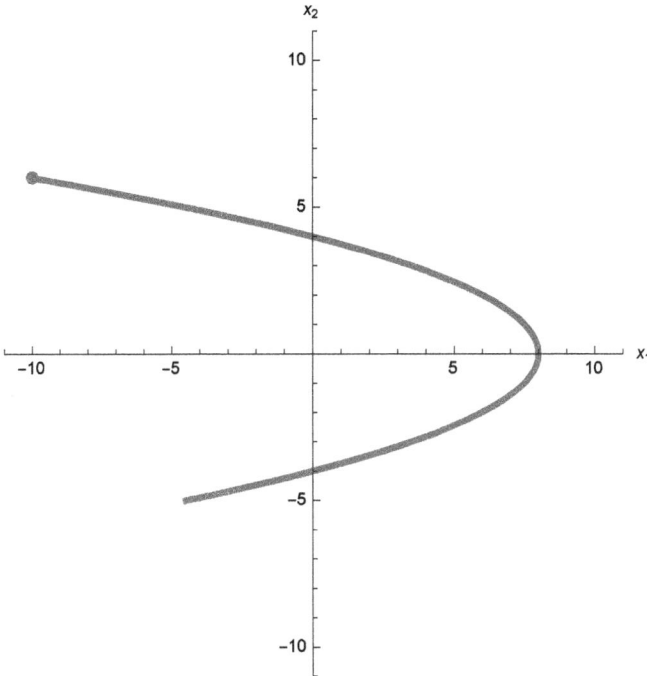

**Figure 11.2:** Position–velocity plane for nanosatellite.

### 11.2.1 Analyzing multiple state equations with Python or Mathematica

We can solve multiple state equations simultaneously to determine formulas for state trajectories via the dsolve or DSolve commands as in Section 9.2.4, and we can use the streamplot or StreamPlot commands to generate "streamplots" showing a selection of state trajectories. We will illustrate how to do this in Python and Mathematica for the following pair of state equations:

$$\dot{x}_1(t) = x_2(t),$$
$$\dot{x}_2(t) = 1.$$

(11.2)

**Python**
We can solve (11.2) via the Python commands

```
#import functions and define variables
from sympy import symbols, Function, dsolve, Eq, diff
t = symbols("t")
x1 = Function('x1')
x2 = Function('x2')
```

```
#solve the system
print(dsolve([Eq(diff(x1(t),t),x2(t)),Eq(diff(x2(t),t),1)]))
```

to obtain

```
[Eq(x1(t), C1 + C2*t + t**2/2), Eq(x2(t), C2 + t)]
```

which translates to the state trajectory pair $x_1(t) = c_1 + c_2 t + \frac{t^2}{2}$ and $x_2(t) = c_2 + t$ with arbitrary constants $c_1$ and $c_2$.

We can also generate a streamplot showing these same state trajectories through a variety of points in the $x_1 x_2$-plane via the following commands:

```
#import commands
from numpy import mgrid
from matplotlib.pyplot import streamplot


#build sample set
x2, x1 = mgrid[-3:3:100j, -3:3:100j]


#streamplot
streamplot(x1, x2, 0*x1+1*x2+0, 0*x1+0*x2+1)
```

Notice that we included the "missing" terms in the system multiplied by zero in the arguments of the streamplot command, since that ensures that the command will recognize the system as having the same dimension as the sample set. It is often useful to overlay the streamplot of another system. For instance, the streamplot for the system

$$\dot{x}_1(t) = x_2(t),$$
$$\dot{x}_2(t) = -1 \tag{11.3}$$

can be overlayed on the one we already generated simply by adding the command

```
streamplot(x1, x2, 0*x1+1*x2+0, 0*x1+0*x2-1)
```

to the preceding list of commands. Python will automatically use different colors for the streamplots of each system, so you can tell them apart.

**Mathematica**

Mathematica's DSolve command solves (11.2) via the command

```
DSolve[{x1'[t] == x2[t], x2'[t] == 1}, {x1[t], x2[t]}, t]
```

to obtain

```
{x1[t] -> t^2/2 + C[1] + t C[2], x2[t] -> t + C[2]}
```

which translates to the state trajectory pair $x_1(t) = \frac{t^2}{2} + c_1 + c_2 t$ and $x_2(t) = t + c_2$ with arbitrary constants $c_1$ and $c_2$.

Mathematica can generate a streamplot for this same system (11.2) via the command

```
StreamPlot[{x2, 1}, {x1, -3, 3}, {x2, -3, 3}]
```

and can generate an overlay on this of a streamplot for the system (11.3) via the command

```
StreamPlot[{{x2, 1}, {x2, -1}}, {x1, -3, 3}, {x2, -3, 3},
    StreamColorFunction -> None]
```

where the extra option at the end of the command ensures different colors for the two different streamplots.

### Lab Activity 4.

1.  *Use overlayed streamplots corresponding to the controls $u^* = 1$ and $u^* = -1$ to analyze the state trajectories for the nanosatellite corresponding to various initial states $(a, b)$.*

    i)   *Identify all the initial states $(a, b)$ that can be optimally controlled to the origin without switching the control.*

    ii)  *Identify the switching curve in the $x_1x_2$-plane where optimal state trajectories switch from one optimal control extreme to the other.*

    iii) *Identify all the initial states $(a, b)$ that can be optimally controlled to the origin by switching $u^*$ from $1 \to -1$.*

    iv)  *Identify all the initial states $(a, b)$ that can be optimally controlled to the origin by switching $u^*$ from $-1 \to 1$.*

2.  *Express the switching curve in parametric form*

    $$sc_1(s) = \ ?,$$
    $$sc_2(s) = \ ?,$$

    *with $sc_1$ representing the $x_1$-coordinates of the switching curve and $sc_2$ representing its $x_2$-coordinates. (Hint: use a point that you are sure the switching curve goes through as the "initial" point and interpret the solutions with negative time s.)*

## Exercises

1.  For each of the following initial states, use the formulas for the optimal state trajectories and the parametric form of the switching curve to find the final elapsed time and the switching time for the time-optimal nanosatellite controlled to position 0 at rest:

    a)  $(a, b) = (-8, 4)$.
    b)  $(a, b) = (10, -8)$.
    c)  $(a, b) = (10, 8)$.

2.  Use the motion of the nanosatellite to explain why the largest final elapsed time in the preceding question should be larger than the others.

# 12 Real-eigenvalue systems

In the preceding chapter, we considered time-optimal control of a simple nanosatellite. Now we will expand that study to the time-optimal control of a pair of states governed by any linear state equation whose defining matrix has real eigenvalues. The same approach is easily generalized to more states.

## 12.1 Linear state equation

The general linear state equation associated with a single control $u$ and states $x_1$ and $x_2$ has the form

$$\underbrace{\begin{bmatrix} \dot{x}_1 \\ \dot{x}_2 \end{bmatrix}}_{\dot{\mathbf{x}}} = \underbrace{\begin{bmatrix} a_{1,1} & a_{1,2} \\ a_{2,1} & a_{2,2} \end{bmatrix}}_{A} \cdot \underbrace{\begin{bmatrix} x_1 \\ x_2 \end{bmatrix}}_{\mathbf{x}} + u \underbrace{\begin{bmatrix} b_1 \\ b_2 \end{bmatrix}}_{\mathbf{b}} \tag{12.1}$$

for some fixed matrix $A$ and vector $\mathbf{b}$. For instance, the nanosatellite state equation has

$$A = \begin{bmatrix} 0 & 1 \\ 0 & 0 \end{bmatrix} \quad \text{and} \quad \mathbf{b} = \begin{bmatrix} 0 \\ 1 \end{bmatrix}.$$

### 12.1.1 Time-optimal control

Suppose the system is to be controlled time-optimally via a bounded control $|u(t)| \leq 1$, Exercise 12-1 confirms that the costate equation is

$$\dot{\lambda} = -A^T \cdot \lambda \tag{12.2}$$

in terms of the transpose of the matrix $A$,

$$A^T = \begin{bmatrix} a_{1,1} & a_{2,1} \\ a_{1,2} & a_{2,2} \end{bmatrix}.$$

Exercise 12-2 confirms that the optimal control from Pontryagin's principle is

$$u^* = \text{sgn}(b_1\lambda_1(t) + b_2\lambda_2(t)) \tag{12.3}$$

in terms of the switching function $s(t) = b_1\lambda_1(t) + b_2\lambda_2(t)$.

https://doi.org/10.1515/9783111290157-012

## 12.1.2 Equilibria

As long as the matrix $A$ is nonsingular (i. e., $\det(A) \neq 0$), the state equation has one equilibrium solution (when $\dot{x} = 0$) at

$$x = -uA^{-1} \cdot b. \tag{12.4}$$

The trajectories for the controlled state system look just like the trajectories for the "unperturbed" ($u = 0$) system $\dot{x} = A \cdot x$ but with the origin shifted to the equilibrium solution (12.4).

## 12.1.3 Two distinct real eigenvalues

When $A$ has distinct real eigenvalues $\epsilon_1 \neq \epsilon_2$, the equilibrium is a source, sink, or saddle, depending on the signs of the eigenvalues. In this case, the transpose matrix $A^T$ has exactly the same eigenvalues $\epsilon_1$ and $\epsilon_2$ as $A$, so it follows from the costate equation (12.2) that the costate vector has the form

$$\lambda(t) = c_1 e^{-\epsilon_1 t} \mathbf{v}_1 + c_2 e^{-\epsilon_2 t} \mathbf{v}_2 = \begin{bmatrix} c_1 e^{-\epsilon_1 t} v_{1,1} + c_2 e^{-\epsilon_2 t} v_{2,1} \\ c_1 e^{-\epsilon_1 t} v_{1,2} + c_2 e^{-\epsilon_2 t} v_{2,2} \end{bmatrix}$$

in terms of the eigenvectors

$$\mathbf{v}_1 = \begin{bmatrix} v_{1,1} \\ v_{1,2} \end{bmatrix} \quad \text{and} \quad \mathbf{v}_2 = \begin{bmatrix} v_{2,1} \\ v_{2,2} \end{bmatrix}$$

of $-A^T$. We can then use this, together with $s(t) = b_1 \lambda_1(t) + b_2 \lambda_2(t)$, to solve for the switching function

$$\begin{aligned} s(t) &= b_1 \lambda_1(t) + b_2 \lambda_2(t) \\ &= b_1 \left( c_1 e^{-\epsilon_1 t} v_{1,1} + c_2 e^{-\epsilon_2 t} v_{2,1} \right) + b_2 \left( c_1 e^{-\epsilon_1 t} v_{1,2} + c_2 e^{-\epsilon_2 t} v_{2,2} \right) \\ &= c_1 (b_1 v_{1,1} + b_2 v_{1,2}) e^{-\epsilon_1 t} + c_2 (b_1 v_{2,1} + b_2 v_{2,2}) e^{-\epsilon_2 t} \\ &= c_3 e^{-\epsilon_1 t} + c_4 e^{-\epsilon_2 t} \end{aligned} \tag{12.5}$$

in terms of the constants $c_3 = c_1(\mathbf{b} \cdot \mathbf{v}_1)$ and $c_4 = c_2(\mathbf{b} \cdot \mathbf{v}_2)$ computed via the dot-products of $\mathbf{b}$ with $\mathbf{v}_1$ and $\mathbf{v}_2$, respectively.

**Problem 12.1.** *Determine how many switches can occur for the switching function* (12.5).

**Lab Activity 5.** *Investigate the time-optimal state trajectories for each of the following two systems to be controlled to* $(0,0)$ *in the* $x_1 x_2$*-plane:*

$$(\text{i}) \quad \dot{x} = \begin{bmatrix} -3 & 2 \\ 2 & -3 \end{bmatrix} x + u \begin{bmatrix} 5 \\ 0 \end{bmatrix} \quad \& \quad (\text{ii}) \quad \dot{x} = \begin{bmatrix} 3 & 2 \\ 2 & 3 \end{bmatrix} x + u \begin{bmatrix} 5 \\ 0 \end{bmatrix}.$$

1.  *Sketch the switching curve in each case and indicate which regions of the $x_1x_2$-plane have which optimal controls.*
2.  *Find the equilibria associated with each system (and using each optimal control) and then find the eigenvalues of the system matrix to determine whether each equilibrium is a sink, source, or saddle.*

## Exercises

12-1.  Use the Hamiltonian to derive the costate equation (12.2) for the two-state case.
12-2.  Use Pontryagin's principle to derive the general optimal control (12.3) for the two-state case.
12-3.  Express the switching curves associated with each system from Lab Activity 5 in parametric form and use these expressions to determine the endpoints (if any) in the $x_1x_2$-plane of each switching curve.

# 13 Controllability

In this chapter, we will address the issue of which initial states can be optimally controlled to a given target. We will revisit the two problems from Lab Activity 5 with this issue in mind.

## 13.1 Lab recap

In Lab Activity 5, we studied the state trajectories associated with the following two systems:

$$\text{(i)} \quad \dot{\mathbf{x}} = \begin{bmatrix} -3 & 2 \\ 2 & -3 \end{bmatrix} \mathbf{x} + u \begin{bmatrix} 5 \\ 0 \end{bmatrix} \quad \& \quad \text{(ii)} \quad \dot{\mathbf{x}} = \begin{bmatrix} 3 & 2 \\ 2 & 3 \end{bmatrix} \mathbf{x} + u \begin{bmatrix} 5 \\ 0 \end{bmatrix}$$

to be controlled as quickly as possible to $(0,0)$ in the $x_1 x_2$-plane.

### 13.1.1 System (i)

The unperturbed ($u = 0$) system has a single equilibrium at $(0,0)$, and this equilibrium is a sink since the eigenvalues of the matrix are both negative ($-5$ and $-1$). Figure 13.1 illustrates this situation, where we can see that any initial state approaches $(0,0)$ in this system as time passes. It happens that this approach slows down for states near the target $(0,0)$ and that the trajectories actually never quite get there (in finite time). Thus nonzero controls are required to get to the target.

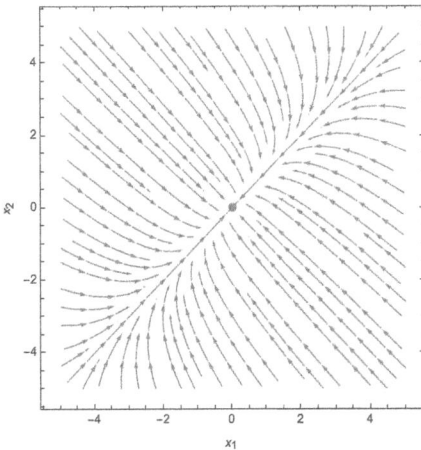

**Figure 13.1:** The unperturbed system.

https://doi.org/10.1515/9783111290157-013

The optimal controls $u^* = \pm 1$ shift the equilibrium to a new location: $(3, 2)$ for $u^* = 1$ and $(-3, -2)$ for $u^* = -1$, as shown in Figure 13.2; otherwise, the trajectories are unchanged. The special trajectories that pass through $(0, 0)$ make up the switching curve as shown in Figure 13.3. There we can also see that every initial state can be optimally controlled to $(0, 0)$, first with $u^* = 1$ (if below the switching curve) or $u^* = -1$ (if above the switching curve) and along a trajectory until it meets the opposite-control part of the switching curve, then the control can be switched, and the trajectory follows the switching curve to $(0, 0)$.

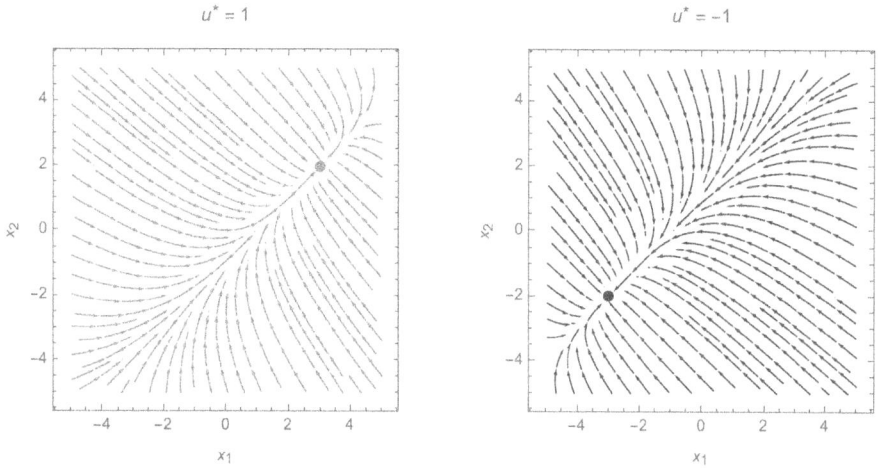

**Figure 13.2:** Optimally controlled systems.

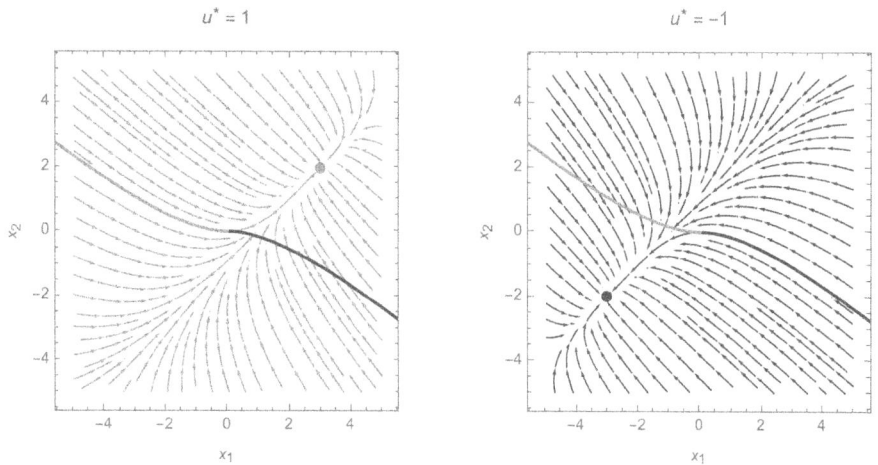

**Figure 13.3:** Switching curve.

### 13.1.2 System (ii)

The unperturbed ($u = 0$) system has a single equilibrium at $(0, 0)$, and this equilibrium is a source since the eigenvalues of the matrix are both positive (5 and 1). Figure 13.4 illustrates this situation, where we can see that any initial state (other than $(0, 0)$) tends away from $(0, 0)$ in this system. Again, nonzero controls are required to get to the target.

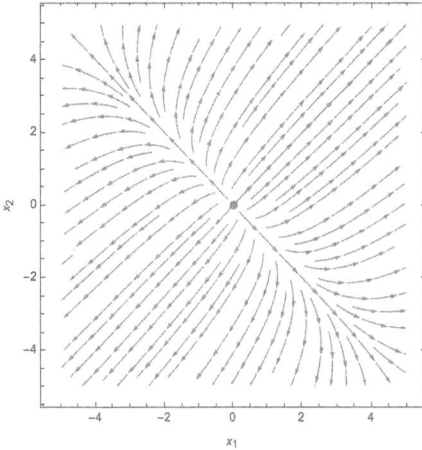

**Figure 13.4:** Unperturbed system.

The optimal controls $u^* = \pm 1$ shift the equilibrium to a new location: $(-3, 2)$ for $u^* = 1$ and $(3, -2)$ for $u^* = -1$, as shown in Figure 13.5; otherwise, the trajectories are unchanged. The special trajectories that pass through $(0, 0)$ again make up the switching curve shown in Figure 13.6, but not all initial states can be controlled to $(0, 0)$ in this case, only those in the *region of controllability* as pictured in Figure 13.7. Every initial state within this region can be optimally controlled to $(0, 0)$, first with $u^* = 1$ (if below the switching curve) or $u^* = -1$ (if above the switching curve) and along a trajectory until it meets the opposite-control part of the switching curve, then the control is switched, and the trajectory follows the switching curve to $(0, 0)$.

On the other hand, initial states outside the region of controllability are unable to counteract the tendency of the unperturbed system to evolve away from the target.

### 13.1.3 Plotting state trajectories with Python or Mathematica

It can be helpful to plot individual state trajectories from a streamplot, and Python or Mathematica have commands to do just this. We illustrate the process for the state trajectory given by

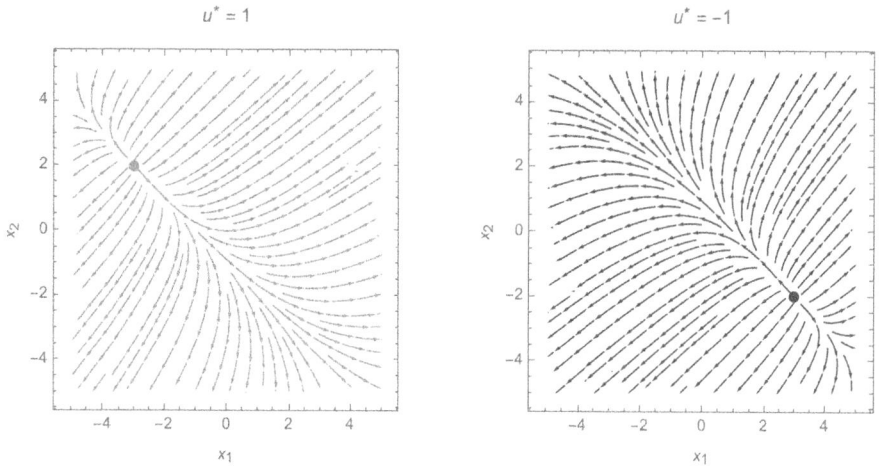

**Figure 13.5:** Optimally controlled systems.

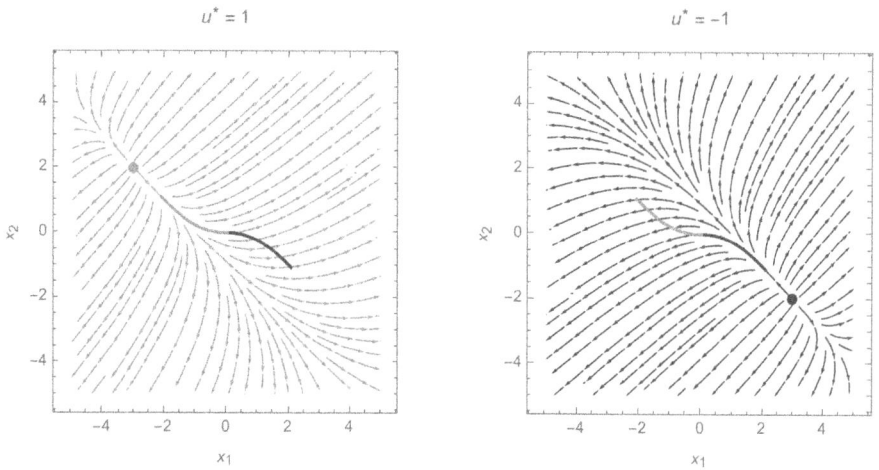

**Figure 13.6:** Switching curve.

$$x_1(t) = \cos(t) - 1,$$
$$x_2(t) = -\sin(t). \tag{13.1}$$

**Python**

The state trajectory given by (13.1) can be plotted for parameters $t$ satisfying $-\pi \le t \le 0$ via the following Python commands:

```
#import functions and define variables
from sympy import symbols, cos, sin, pi
```

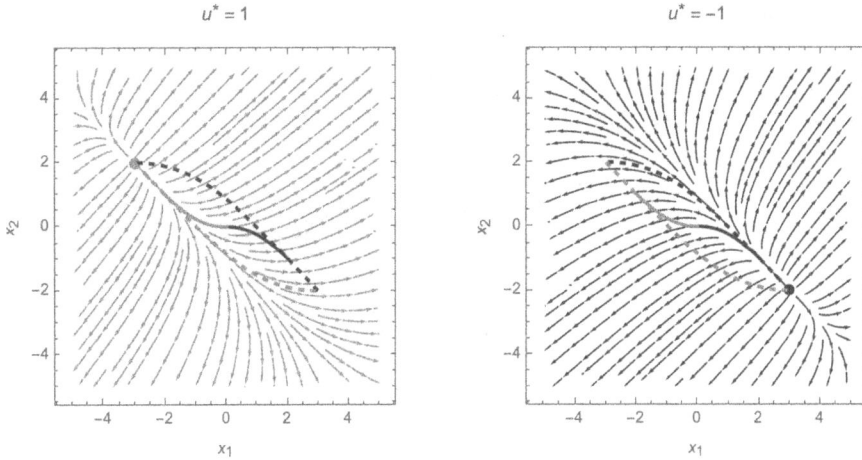

**Figure 13.7:** Region of controllability.

```
from sympy.plotting import plot_parametric
t = symbols('t')

#plot trajectory
plot_parametric(cos(t)-1,-sin(t),(t,-pi,0))
```

**Mathematica**

The state trajectory given by (13.1) can be plotted for parameters $t$ satisfying $-\pi \le t \le 0$ via the following Mathematica command:

```
ParametricPlot[{Cos[t] - 1, -Sin[t]}, {t, -Pi, 0}]
```

**Lab Activity 6.**

1. *Classify the equilibria and identify (both graphically and with formulas) the time-optimal switching curve to $(0,0)$ and the region of controllability for*

$$\dot{\mathbf{x}} = \begin{bmatrix} 1 & 3 \\ 3 & 1 \end{bmatrix} \mathbf{x} + u \begin{bmatrix} -7 \\ -5 \end{bmatrix}.$$

2. *Classify the equilibria and identify (both graphically and with formulas) the time-optimal switching curve to $(0,0)$ and the region of controllability for the nanosatellite*

$$\dot{\mathbf{x}} = \begin{bmatrix} 0 & 1 \\ 0 & 0 \end{bmatrix} \mathbf{x} + u \begin{bmatrix} 0 \\ 1 \end{bmatrix}$$

*with one broken thruster*

$$-1 \le u(t) \le 0.$$

## Exercises

13-1. How long does it take for the unperturbed system

$$\dot{\mathbf{x}} = \begin{bmatrix} -3 & 2 \\ 2 & -3 \end{bmatrix} \mathbf{x}$$

to move the initial state $(0, -5)$ to $(0, 0)$? Compare this to the amount of time it takes to time-optimally control the same initial state to $(0, 0)$ for the (perturbed) system

$$\dot{\mathbf{x}} = \begin{bmatrix} -3 & 2 \\ 2 & -3 \end{bmatrix} \mathbf{x} + u \begin{bmatrix} 5 \\ 0 \end{bmatrix}$$

with $|u(t)| \leq 1$.

13-2. How many switches can occur in the time-optimal control of the system

$$\dot{\mathbf{x}} = \begin{bmatrix} 1 & 0 \\ 1 & 0 \end{bmatrix} \mathbf{x} + u \begin{bmatrix} 1 \\ -1 \end{bmatrix}$$

with $|u(t)| \leq 1$? Fully justify your answer.

13-3. Classify any equilibria (unperturbed or optimally controlled) and identify (both graphically and with formulas) the time-optimal switching curve to $(0, 0)$ and the region of controllability for Exercise 13-2.

# 14 Complex-eigenvalue systems

We have already seen that there is at most one switch in the time-optimal control to $(0,0)$ in the $x_1x_2$-plane when the two states $x_1$ and $x_2$ are determined by a linear state equation whose defining matrix has two distinct real eigenvalues. Now we consider linear state equations whose defining matrices have complex eigenvalues.

## 14.1 Complex-eigenvalue systems

We focus on a prototypical example (adapted from [2, Example 5.6]): the time-optimal control to $(0,0)$ of a two-state vector $\mathbf{x}$ satisfying the state equation

$$\dot{\mathbf{x}} = \underbrace{\begin{bmatrix} a & 1 \\ -1 & a \end{bmatrix}}_{A} \mathbf{x} + u \underbrace{\begin{bmatrix} -a \\ 1 \end{bmatrix}}_{\mathbf{b}},$$

where $a$ is a real number, and the control $u$ satisfies $|u(t)| \le 1$. The matrix $A$ in this case has complex eigenvalues $a \pm i$.

### 14.1.1 Switching

Using our general work for the linear state equations from Chapter 12, we have the costate equation (12.2)

$$\dot{\lambda} = -A^T \cdot \lambda = \begin{bmatrix} -a & 1 \\ -1 & -a \end{bmatrix} \cdot \lambda$$

and optimal control (12.3)

$$u^* = \text{sgn}(s(t)) \text{ for } s(t) = b_1\lambda_1(t) + b_2\lambda_2(t) = -a\lambda_1(t) + \lambda_2(t).$$

By solving the costate equation we get the costates

$$\lambda_1(t) = e^{-at}(c_1 \cos(t) + c_2 \sin(t)),$$
$$\lambda_2(t) = e^{-at}(c_2 \cos(t) - c_1 \sin(t)),$$

so that our optimal control is the sign of the trigonometric switching function

$$\begin{aligned} s(t) &= -a(e^{-at}(c_1 \cos(t) + c_2 \sin(t))) + e^{-at}(c_2 \cos(t) - c_1 \sin(t)) \\ &= e^{-at}((-ac_1 + c_2) \cos(t) + (-ac_2 - c_1) \sin(t)) \\ &= c_3 e^{-at} \sin(t + c_4), \end{aligned}$$

https://doi.org/10.1515/9783111290157-014

where we have applied the identity

$$a \cos(t) + b \sin(t) = \pm \sqrt{a^2 + b^2} \sin\left(t + \tan^{-1}\frac{a}{b}\right)$$

with $a = -ac_1 + c_2$ and $b = -ac_2 - c_1$ to get the final line. From this description we see that the switching function has infinitely many zeroes spaced exactly $\pi$ units apart in $t$. We conclude that:

- The optimal control $u^*$ can have multiple switches (between its extremes 1 and $-1$);
- The first switch happens after no more than $\pi$ time units have elapsed, and
- Each subsequent switch occurs after precisely $\pi$ units of time have elapsed since the preceding switch.

**Lab Activity 7.** *Identify the switching curve for the time-optimal control of the system*

$$\dot{\mathbf{x}} = \begin{bmatrix} a & 1 \\ -1 & a \end{bmatrix} \mathbf{x} + \begin{bmatrix} -a \\ 1 \end{bmatrix} u$$

*to $(0, 0)$ with $|u(t)| \le 1$ for the following values of $a$:*

1) $a = 0$
2) $a = -0.2$
3) $a = 0.2$

## Exercises

14-1. Use appropriate parameterizations to find the optimal time for the time-optimal control to $(0, 0)$ from $(3, 3)$ of the system

$$\dot{\mathbf{x}} = \begin{bmatrix} 0 & 1 \\ -1 & 0 \end{bmatrix} \mathbf{x} + u \begin{bmatrix} 0 \\ 1 \end{bmatrix}$$

with $|u(t)| \le 1$.

(Hint: solve for $x_2$ in terms of $x_1$ from the formula for the appropriate half-circle on the switching curve.)

14-2. Use appropriate parameterizations to find the optimal time in the same control problem as in Exercise 14-1 but instead with $0 \le u(t) \le 1$.

14-3. Explain the switching possible for the time-optimal control of the system

$$\dot{\mathbf{x}} = \begin{bmatrix} 0 & 1 \\ -1 & 0 \end{bmatrix} \mathbf{x} + \begin{bmatrix} u_1 \\ u_2 \end{bmatrix}$$

with two controls satisfying $|u_1(t)| \le 1$ and $|u_2(t)| \le 1$. You may want to make use of the identities $a \cos(t) + b \sin(t) = \pm\sqrt{a^2 + b^2} \sin(t + \tan^{-1}\frac{a}{b})$ and $\tan^{-1}(\frac{-b}{a}) = \tan^{-1}(\frac{a}{b}) - \frac{\pi}{2}$.

# 15 Controllability (complex eigenvalues)

In the preceding chapter, we produced switching curves associated with three linear systems having complex eigenvalues. Now we consider the regions of controllability for these systems.

## 15.1 Lab recap

In Lab Activity 7, we looked at the time-optimal state trajectories associated with the system

$$\dot{\mathbf{x}} = \begin{bmatrix} a & 1 \\ -1 & a \end{bmatrix} \mathbf{x} + \begin{bmatrix} -a \\ 1 \end{bmatrix} u$$

with $|u(t)| \leq 1$ for three different values of $a$.

1. $a = 0$: The switching curve in this case is pictured in Figure 15.1 and continues in the same pattern along the $x_1$-axis. Notice that only the semicircles through the origin $(0,0)$ represent trajectories of the system. The other semicircles are simply locations in the $x_1 x_2$-plane where a switch is made.

**Question 15.1.** *What is the region of controllability in this case?*

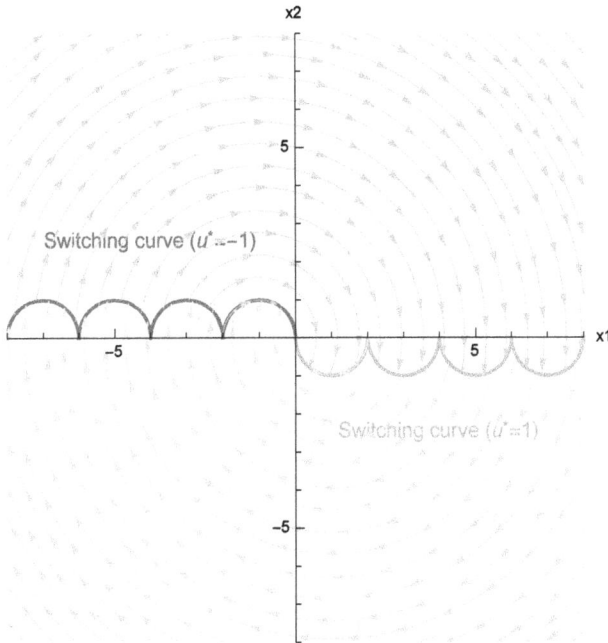

**Figure 15.1:** Switching curve for $a = 0$.

https://doi.org/10.1515/9783111290157-015

2. $a = -0.2$: The switching curve in this case is pictured in Figure 15.2 and continues in the same pattern along the $x_1$-axis (expanding in size).

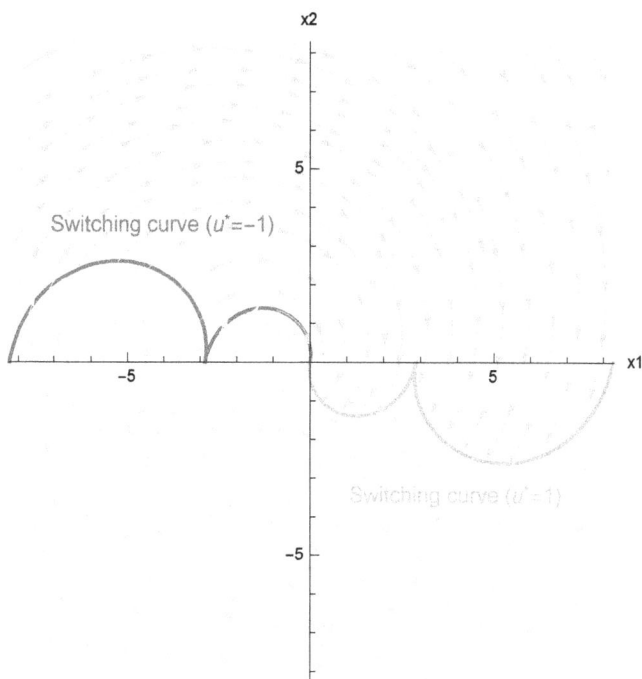

**Figure 15.2:** Switching curve for $a = -0.2$.

**Question 15.2.** *What is the region of controllability in this case?*

3. $a = 0.2$: The switching curve in this case is pictured in Figure 15.3 and continues in the same pattern along the $x_1$-axis (shrinking in size).

**Question 15.3.** *What is the region of controllability in this case?*

## 15.2 Region of controllability for case $a = 0.2$

For more precision on the switching curve in the case $a = 0.2$, we solve the state equation for the optimal trajectories associated with this system:

$$x_1^*(t) = e^{0.2t}(c_1 \cos(t) + c_2 \sin(t)) + u^*,$$
$$x_2^*(t) = e^{0.2t}(c_2 \cos(t) - c_1 \sin(t)), \tag{15.1}$$

where we have used the fact that the optimal control $u^*$ is effectively constant.

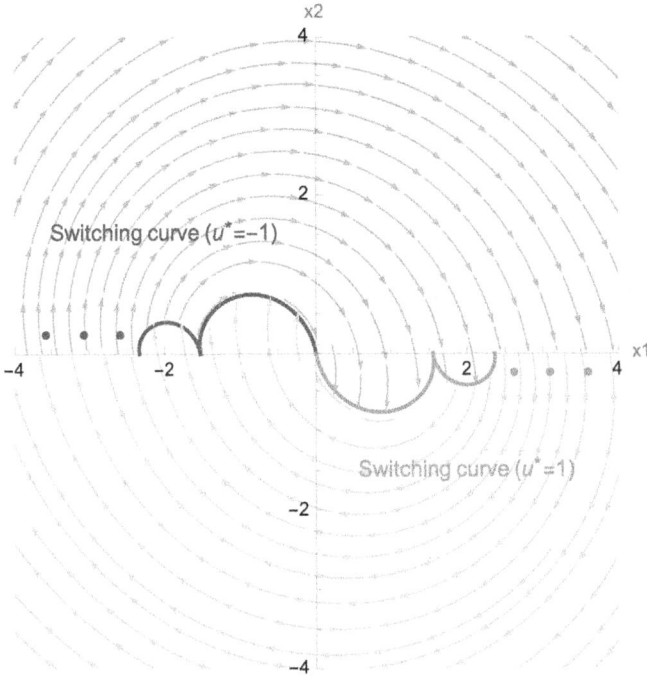

**Figure 15.3:** Switching curve for $a = 0.2$.

We label the endpoints on the $x_1$-axis of each part of the switching curve as shown in Figure 15.4 and we want to determine the exact values of $m_1, p_1, m_2, p_2$, and so on ($m$ for the "minus" side of the $x_1$-axis and $p$ for "plus" side). If we can show that the values $m_n$ and $p_n$ converge as $n \to \infty$, then we know that the switching curve (and consequently the region of controllability) is bounded.

### 15.2.1 Determining exact values of $m_1$, $p_1$, and $m_2$

We know that $m_1 = x_1^*(0)$ is the initial $x_1$-coordinate of the optimal state trajectory that follows optimal control $u^* = -1$ to $(0, 0)$ in precisely $\pi$ time units. We can thus use (15.1) to get

$$0 = x_1^*(\pi) = c_1 e^{0.2\pi} \cos(\pi) - 1 \quad \Longrightarrow \quad c_1 = -e^{-0.2\pi}.$$

Substituting this back into (15.1) and solving for $m_1 = x_1^*(0)$ give us the exact value we seek:

$$m_1 = x_1^*(0) = -e^{-0.2\pi} - 1. \tag{15.2}$$

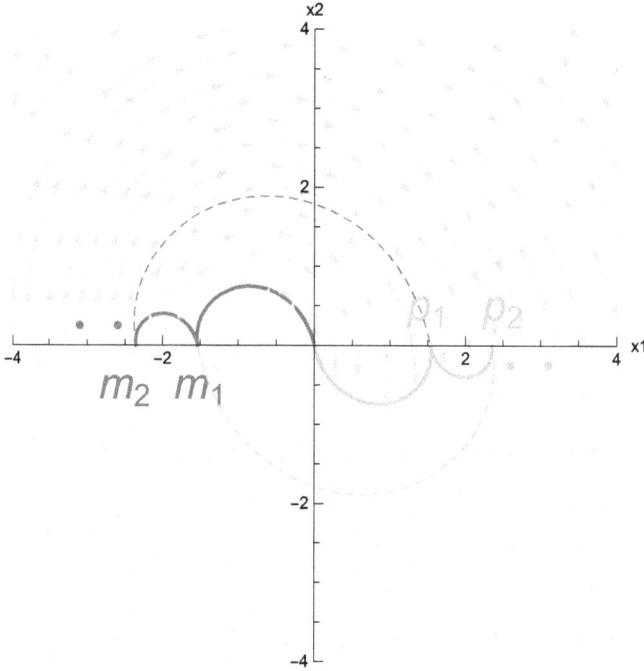

**Figure 15.4:** Labeled switching curve for $a = 0.2$.

Note that this exact value is more useful for our present purposes than a decimal (approximate) value.

**Problem 15.1.**
(a) *Find the exact value of $p_1$.*
(b) *Find the exact value of $m_2$. (Hint: use how $m_2$ is related to $p_1$ as in Figure 15.4.)*

### 15.2.2 Determining the $x_1$-length of the switching curve

Combining the formulas (15.2) and $m_2 = -e^{(-0.2\pi)(2)} - 2e^{-0.2\pi} - 1$, we get

$$m_2 - m_1 = -e^{(-0.2\pi)(2)} - 2e^{-0.2\pi} - 1 - (-e^{-0.2\pi} - 1)$$
$$= -e^{(-0.2\pi)(2)} - e^{-0.2\pi}$$
$$= -e^{-0.2\pi}(1 + e^{-0.2\pi}).$$

In a similar manner, we can determine the exact value of the difference of any consecutive $m_{n+1}$ and $m_n$:

$$m_{n+1} - m_n = -e^{-0.2\pi n}(1 + e^{-0.2\pi}).$$

We can use this to express $m_{n+1}$ as follows:

$$m_{n+1} = m_1 + (m_2 - m_1) + \cdots + (m_{n+1} - m_n)$$

$$= -(1 + e^{-0.2\pi}) \sum_{j=0}^{n} e^{-0.2\pi j}.$$

To find how far to the left along the $x_1$-axis the entire $u^* = -1$ part of the switching curve stretches, we take the limit as $n \to \infty$:

$$\lim_{n\to\infty} m_{n+1} = \lim_{n\to\infty} -(1 + e^{-0.2\pi}) \sum_{j=0}^{n} e^{-0.2\pi j}$$

$$= -(1 + e^{-0.2\pi}) \frac{1}{1 - e^{-0.2\pi}}$$

$$\approx -3.28714$$

(where we applied the geometric series formula $\sum_{j=0}^{\infty} a^j = \frac{1}{1-a}$ to get the second equality). Because of the symmetry in this system, it is easy to see that the entire $u^* = 1$ part of the switching curve stretches the same amount $\approx 3.28714$ to the right along the $x_1$-axis.

### 15.2.3 Bounded region of controllability

We can see in Figure 15.5 that the region of controllability is bounded, and we obtained the boundary by backtracking from the $x_1$-endpoints of the switching curve.

## Exercises

15-1. [2, Exercise 5.2.2] Consider the time-optimal control to $(0, 0)$ from $(a, b)$ for the system

$$\dot{\mathbf{x}} = \begin{bmatrix} 0 & 1 \\ -1 & 0 \end{bmatrix} \mathbf{x} + u \begin{bmatrix} 0 \\ 1 \end{bmatrix}$$

with $|u(t)| \leq 1$. If $(a, b)$ is in the first quadrant of the $x_1 x_2$-plane, then show that the optimal control has $n$ switches, where $n$ satisfies the inequalities

$$2n - 1 \leq \sqrt{(a + 1)^2 + b^2} \leq 2n + 1.$$

15-2. [3, adapted Exercise 18.7] Consider the time-optimal control to $(0, 0)$ from $(a, b)$ for the (nonlinear) system

$$\dot{x}_1 = e^{x_2},$$

$$\dot{x}_2 = u$$

with $|u(t)| \leq 1$.

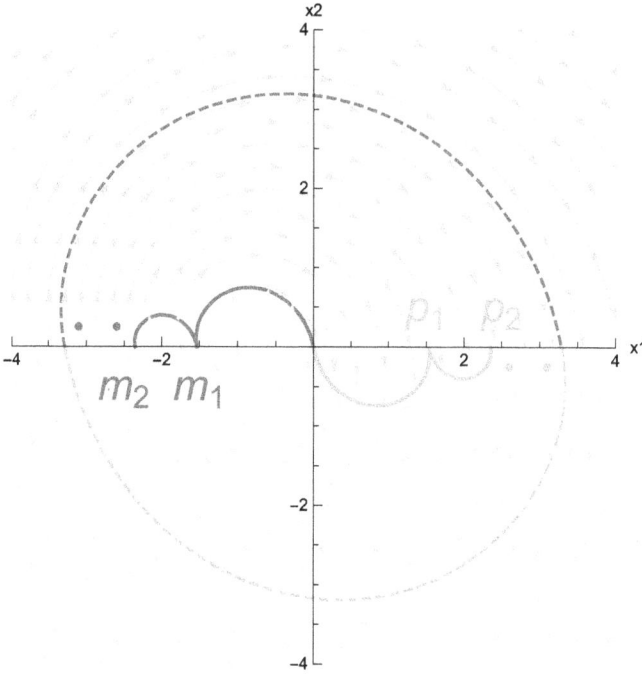

**Figure 15.5:** Region of controllability for $a = 0.2$.

(a) Explain why the optimal control $u^*$ can have at most one switch.
(Hint: without solving the second costate equation, use it to deduce a relevant property of the second costate.)

(b) Give the parametric formulas $x_1^*(t)$ and $x_2^*(t)$ in terms of $u^*$ for the piece of the optimal trajectory starting from $(a, b)$.

(c) Give a formula for the switching curve in terms of the parameter $s \in (-\infty, 0]$.

(d) Give a formula for the region of controllability.

# 16 Control to target curve

We consider optimal control problems where the aim is to control the state to some target curve $x(T) = \text{tar}(T)$ and not necessarily just a fixed end state. This adds a new "costate perpendicularity condition" to the condition $H(T^*) = 0$ we derived for the case of a fixed end state.

## 16.1 Costate perpendicularity

Recall from Chapter 9 that for the case of a single state, we apply the transversality condition to the Lagrangian integrand $\lambda \cdot \dot{x} + H(\lambda, x, u)$ in

$$\int_{t_0}^{T} f_0(x, u) + \lambda \cdot (\dot{x} - f(x, u))\, dt = \int_{t_0}^{T} \lambda \cdot \dot{x} + H(\lambda, x, u)\, dt$$

to get

$$\lambda(T^*) \cdot \dot{x}^*(T^*) + H(T^*) + (\text{tar}'(T^*) - \dot{x}^*(T^*)) \cdot \lambda(T^*) = 0.$$

In Chapter 9, this simplified to $H(T^*) = 0$ since $\text{tar}'(T^*) = 0$, but in general it simplifies to

$$H(T^*) + \text{tar}'(T^*) \cdot \lambda(T^*) = 0. \tag{16.1}$$

As long as the target curve is not a vertical line in the $tx$-plane, we can satisfy this equation by ensuring that the following pair of equations hold:

$$H(T^*) = 0 \quad \text{and} \quad \text{tar}'(T^*) \cdot \lambda(T^*) = 0. \tag{16.2}$$

In the special case where the target curve is a vertical line in the $tx$-plane, we instead divide both sides of (16.1) by $\text{tar}'(T^*) = \infty$ to get $\lambda(T^*) = 0$ (with no restriction on $H(T^*)$). The first equation $H(T^*) = 0$ in (16.2) is the usual condition we have already seen, and the second equation $\text{tar}'(T^*) \cdot \lambda(T^*) = 0$ is the *costate perpendicularity condition*.

## 16.2 Two-state case

The title "costate perpendicularity" makes more sense in the two-state case where we have some (parameterized) target curve

$$\mathbf{c}(t) = c_1(t)\vec{\imath} + c_2(t)\vec{\jmath}$$

https://doi.org/10.1515/9783111290157-016

expressed as a position vector with two components $c_1(t)$ and $c_2(t)$, as shown in Figure 16.1. A generalization of the argument above results in the equation $H(T^*) = 0$ paired with the costate perpendicularity condition

$$\dot{c}_1(T^*) \cdot \lambda_1(T^*) + \dot{c}_2(T^*) \cdot \lambda_2(T^*) = 0$$

$$\updownarrow$$

$$\underbrace{\left( \dot{c}_1(T^*)\vec{i} + \dot{c}_2(T^*)\vec{j} \right)}_{\text{tangent to target curve}} \cdot \left( \lambda_1(T^*)\vec{i} + \lambda_2(T^*)\vec{j} \right) = 0.$$

Graphically, this means that if we were to draw the costate vector $\lambda_1(T^*)\vec{i} + \lambda_2(T^*)\vec{j}$ in the $x_1x_2$-plane as in Figure 16.2, it would be perpendicular to the target curve at the point where the optimal trajectory ends.

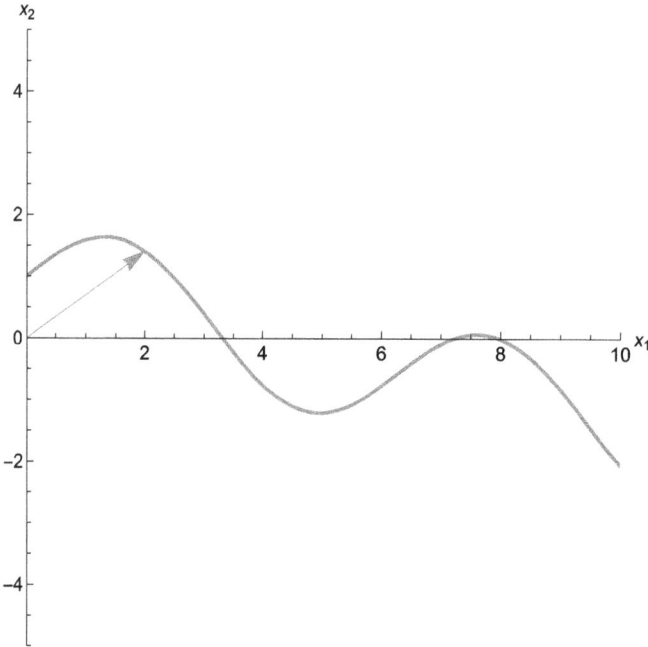

**Figure 16.1:** Target curve $c(t) = c_1(t)\vec{i} + c_2(t)\vec{j}$.

**Example 7** (Nanosatellite). If we want to control our nanosatellite to rest (anywhere) as quickly as possible, then we have the usual state equations

$$\dot{x}_1 = x_2 \quad \text{and} \quad \dot{x}_2 = u$$

and the cost integrand $f_0(x_1, x_2, u) = 1$. As usual, we construct the Hamiltonian for this problem as

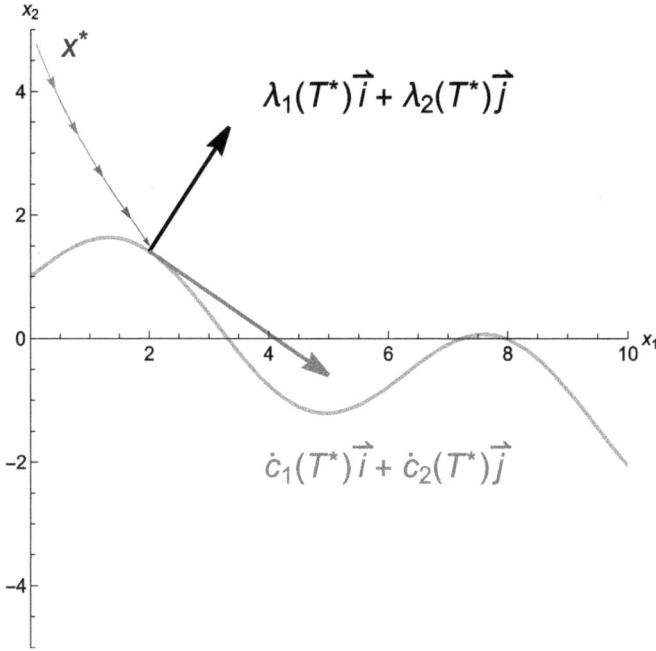

**Figure 16.2:** Costate perpendicularity illustrated.

$$H(\lambda, \mathbf{x}, u) = 1 - \lambda_1 x_2 - \lambda_2 u,$$

which generates the costate equations

$$\dot{\lambda}_1 = 0 \quad \Longrightarrow \quad \lambda_1(t) = c_1,$$
$$\dot{\lambda}_2 = -\lambda_1 \quad \Longrightarrow \quad \lambda_2(t) = c_2 - c_1 t.$$

**Question 16.1.** *What is the target curve in this case?*

**Question 16.2.** *What does costate perpendicularity say in this case?*

We conclude that the costates are $\lambda_1(t) = 0$ and $\lambda_2(t) = c_2$, so that the Hamiltonian simplifies to

$$H(\lambda, \mathbf{x}, u) = 1 - c_2 u.$$

Pontryagin's principle says $u = \mathrm{sgn}(c_2)$, so there are no switches (since $c_2$ is constant).

Figure 16.3 synthesizes the optimal controls on the $x_1 x_2$-plane using the optimal control $u^* = -1$ for initial points above the target curve (the $x_1$-axis) and $u^* = 1$ for initial points below the target curve. In terms of the nanosatellite, this says that we accelerate as much as possible to the left if the initial velocity is moving the nanosatellite to the right, and we accelerate as much as possible to the right if the initial velocity is moving

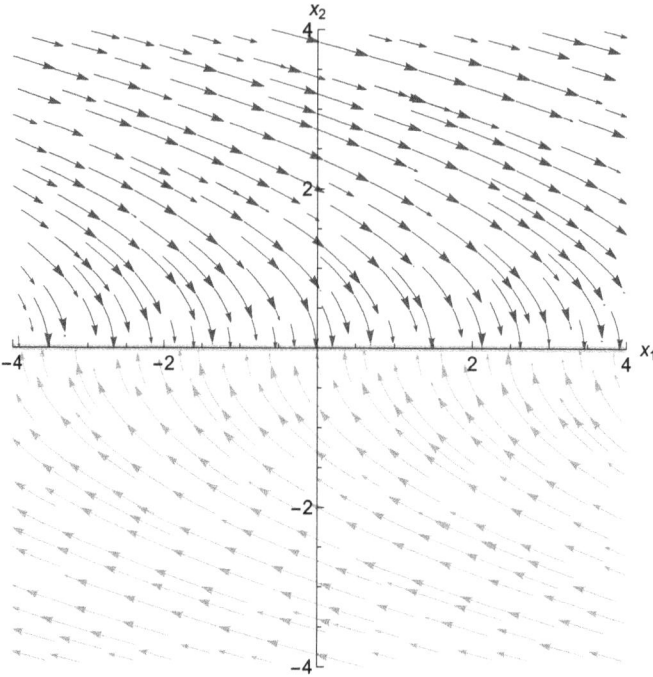

**Figure 16.3:** Optimal control synthesis.

the nanosatellite to the left. Essentially, we slam on the "brakes" if we want to bring the nanosatellite to rest as quickly as possible.

**Problem 16.1.** *Use costate perpendicularity to help you sketch the switching curve and indicate the time-optimal control strategies for the nanosatellite controlled to rest at any nonnegative position.*

## Exercises

16-1. [2, adapted Example 5.9] Investigate the time-optimal state trajectories for the system

$$\dot{\mathbf{x}} = \begin{bmatrix} 0 & 1 \\ -1 & 0 \end{bmatrix} \mathbf{x} + u \begin{bmatrix} 0 \\ 1 \end{bmatrix}$$

controlled to the target line $x_2 = x_1$ with $|u(t)| \le 1$.

(i) Use costate perpendicularity and $H(T^*) = 0$ to show that the target curve cannot be reached above $(0.5, 0.5)$ by an optimal trajectory with $u^*(T^*) = 1$ and cannot be reached below $(-0.5, -0.5)$ by an optimal trajectory with $u^*(T^*) = -1$.

(ii) Now consider a point $(p, p)$ on the target curve and an optimal trajectory arriving there with control $u^*(T^*) = -1$. Use the same conditions (costate perpendicularity and $H(T^*) = 0$) to solve for $c_3$ and $c_4$ (in terms of $p$ and $T^*$) from

$$\lambda_1(t) = c_3 \cos(t) + c_4 \sin(t),$$
$$\lambda_2(t) = c_4 \cos(t) - c_3 \sin(t).$$

(16.3)

(iii) Use your solutions to solve (by hand) for the zeroes of the switching function $s(t)$ in terms of $T^*$ and graph the result as a function of $T^*$ from 0 to 2. You should see the line $t = T^* - b$ of slope one and intercept $-b < 0$.

(iv) Find $b$ and explain why this shows that if the trajectory for $u^* = -1$ arrives at the target at time $T^* > b$, then it must have switched from $u^* = 1$ exactly $b$ time units before $T^*$.

(v) A similar argument shows that if the trajectory for $u^* = 1$ arrives at the target at time $T^* > b$, then it must have switched from $u^* = -1$ exactly $b$ time units before $T^*$. Apply these facts to sketch the switching curve and indicate the optimal control strategies in this case.

16-2. Consider the time-optimal control of the nanosatellite with $|u(t)| \leq 1$.

(a) Sketch the switching curve and indicate the optimal control strategies if the goal is to get the nanosatellite to rest at any position within 5 units of position zero?

(b) Sketch the switching curve and indicate the optimal control strategies if the goal is to get the nanosatellite to position zero and having any speed with magnitude less than or equal to 5?

16-3. For the system $\dot{x} = x + u$ with cost functional

$$\int_0^{\ln 2} x^2 + (x + u)^2 \, dt$$

and unbounded control, find the optimal control in each of the following two cases:
(a) $x(0) = 1$ and $x(\ln 2) = 1$.
(b) $x(0) = 1$.

16-4. For the system $\dot{x} = -x + \sqrt{3}u$ with cost functional

$$\frac{1}{2} \int_0^T x^2 + u^2 \, dt$$

and unbounded control, find the optimal control in each of the following two cases:
(a) $x(0) = 5$ and $x(T) = 4$.
(b) $x(0) = 5$ and $T = 1$.

# 17 Singular controls

We introduce the notion of "singular" controls, which occur when the optimal control is unspecified by Pontryagin's principle. To motivate that introduction, we begin with the following problem.

**Problem 17.1** ([2, adapted Example 5.10]). *Use Pontryagin's principle to find the optimal controls (and time) that move an object in the $x_1 x_2$-plane from position $(a, b)$ to $(0, 0)$ as quickly as possible by separately controlling the velocity (bounded in magnitude by 1) in each coordinate direction.*

## 17.1 Singular controls

Whenever the optimal control is not determined by Pontryagin's principle, we say that it is a *singular control*. When there is a switching function $s(t)$, singular control entails $s(t) = 0$ for an entire subinterval $[t_1, t_2]$ of $[0, T]$. The isolated zeroes of $s(t)$ simply represent the moments when the optimal control switches from one extreme to another and do not correspond to singular controls. For instance, the switching function $s(t) = c_2 - c_1 t$ associated with the optimal control in the nanosatellite problem has at most one isolated zero unless it is identically zero (when $c_1 = c_2 = 0$), and we can rule out that exception since it implies that the Hamiltonian is identically 1 (which clearly violates the condition $H(T^*) = 0$). The example in Problem 17.1 is one where we cannot rule out singular control in this manner.

**Example 8.** In Problem 17.1, we can have either switching function $s_1(t) = c_1$ or $s_2(t) = c_2$ equal to zero (but not both) without violating

$$0 = H(T^*) = 1 - c_1 u_1^*(T^*) - c_2 u_2^*(T^*).$$

Thus there are two singular cases to consider.

**Singular case 1:** $c_1 = 0$ and $c_2 \neq 0$
As before, we get $u_2^* = -\text{sgn}(b)$ and $T^* = |b|$. However, this time, we only know that the control $u_1$ is restricted to values in $[-1, 1]$, so the first state equation can be integrated to yield

$$x_1^*(t) = \int_0^t u_1^*(\tau) \, d\tau + a.$$

Notice that this formula does not give an entirely satisfying solution for $x_1^*$ until we have the optimal control and can compute the integral. In any case the endpoint condition $x_1^*(T^*) = 0$ thus translates into the following restriction on the optimal control $u_1^*$:

https://doi.org/10.1515/9783111290157-017

$$\int_{0}^{|b|} u_1^*(\tau)\, d\tau = -a \tag{17.1}$$

(where we have applied $T^* = |b|$). Since the optimal control values satisfy $|u_1^*(t)| \leq 1$, we know that the integral in (17.1) satisfies

$$-|b| \leq \int_{0}^{|b|} u_1^*(\tau)\, d\tau \leq |b|,$$

from which deduce that (17.1) can only hold when $|a| \leq |b|$.

We conclude that when $|a| \leq |b|$, the end time is $T^* = |b|$, and the optimal controls are $u_2^* = -\text{sgn}(b)$ and any $u_1^*$ satisfying $|u_1^*(t)| \leq 1$ and (17.1).

**Singular case 2: $c_1 \neq 0$ and $c_2 = 0$**

Analogously to the preceding singular case, we deduce that when $|b| \leq |a|$, the end time is $T^* = |a|$, and the optimal controls are $u_1^* = -\text{sgn}(a)$ and any $u_2^*$ satisfying $|u_2^*(t)| \leq 1$ and

$$\int_{0}^{|a|} u_2^*(\tau)\, d\tau = -b. \tag{17.2}$$

Notice that we always have $T^* = \max\{|a|, |b|\}$ and that the singular optimal control cases both include the nonsingular solution we worked out earlier. Figure 17.1 synthesizes the optimal controls in this case, where the unspecified optimal controls in each region are assumed to satisfy equation (17.1) or (17.2), respectively.

**Problem 17.2.** *Show that the nonsingular case of optimal controls $u_1^* = -\text{sgn}(a)$ and $u_2^* = -\text{sgn}(b)$ when $|a| = |b|$ is covered by the optimal control in singular case 2.*

**Problem 17.3.** *Give an example of optimal controls in Problem 17.1 from the initial state $(2, 1)$.*

### 17.1.1 Singular case means $\dot{s}(t) = 0$

One trick that sometimes is useful when investigating the singular case is noticing that when $s(t) = 0$ on an entire subinterval $[t_1, t_2]$ of $[0, T]$, the derivative is also zero $\dot{s}(t) = 0$ on $(t_1, t_2)$ (since $s(t) = 0$ is constant on $[t_1, t_2]$). This is illustrated in Figure 17.2, where a switching function is graphed with its derivative. Note that the same reasoning means that higher-order derivatives of $s(t)$ are also zero on $(t_1, t_2)$.

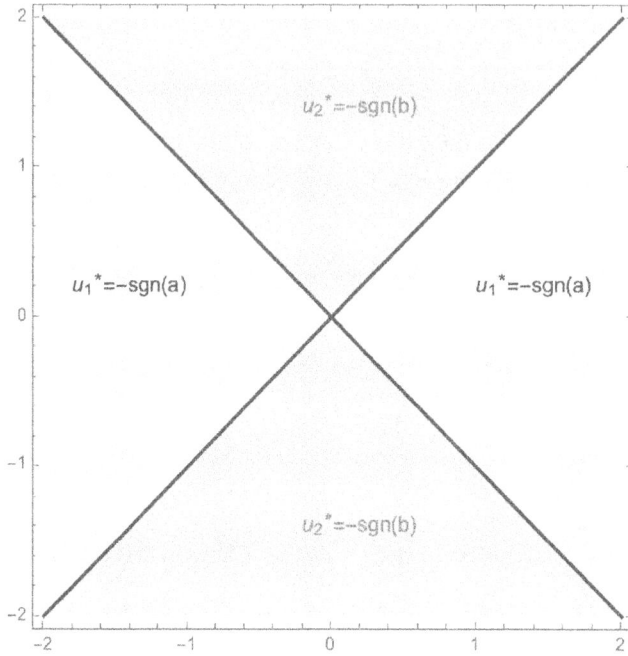

**Figure 17.1:** Optimal control synthesis.

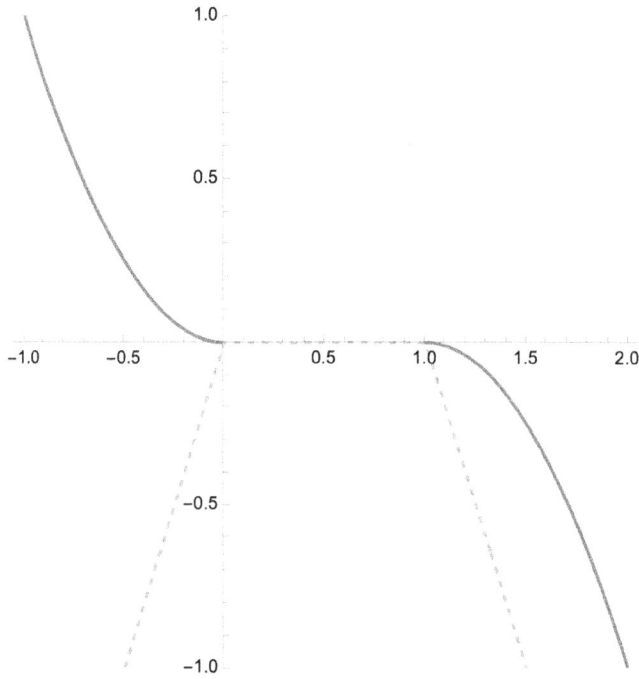

**Figure 17.2:** Switching function $s(t)$ and its derivative $s'(t)$ (dashed).

## Exercises

17-1. [3, adapted Problem E] A simple problem in resource management is to minimize the negative present value

$$\int\limits_0^T e^{-\delta x_2}(\theta - x_1)u\,dt$$

of a resource whose amount $x_1$ changes from $x_1(0) = a$ to $x_1(T) = b$ according to $\dot{x}_1 = x_1(1 - x_1) - qux_1$, where $\delta$, $\theta$, and $q$ are fixed positive constants (here the cost is represented by $\theta u$, revenue is represented by $x_1 u$, and the integral measures the present value of negative profit). The control $u$ represents harvesting effort and satisfies $0 \le u(t) \le 1$.

(a) The second "state" $x_2$ satisfies $x_2(0) = 0$ and $\dot{x}_2 = 1$. What does this state represent and what is the appropriate endpoint condition $x_2(T) = ?$

(b) Determine all possible values for optimal controls.

(Hints: Use $\dot{s}(t) = 0$ in the singular case; do not solve any of the differential equations; manipulate your various equations algebraically until you can solve for $x_1$ explicitly in a form that you can differentiate easily; then solve for $u$ algebraically via the state equation.)

17-2. Determine all possible values for optimal controls of the nanosatellite to be controlled with $|u(t)| \le 1$ while minimizing

$$\int\limits_0^T \frac{x_2^2}{2} - \frac{x_1^2}{2}\,dt.$$

(Hint: use $\dot{s}(t) = 0$ in the singular case.)

# 18 Energy-optimal control

In this chapter, we focus on a particular version of the nanosatellite problem where the goal is (as usual) to move the nanosatellite from position $a$ with initial velocity $b$ to rest at position 0. However, now we are interested in minimizing the energy consumed instead of time.

## 18.1 Energy-optimal control

We have the usual state equations for the nanosatellite

$$\dot{x}_1 = x_2 \quad \text{and} \quad \dot{x}_2 = u,$$

but now the cost functional is $\int_0^T |u| \, dt$ (adapted from [2, Example 5.11]).

**Question 18.1.** *What does this cost functional measure in practical terms?*

**Question 18.2.** *What aspect of a state trajectory in the $x_1 x_2$-plane does this cost functional measure?*

**Question 18.3.** *What is the lower bound on the cost functional $\int_0^T |u| \, dt$ to control $x_2$ from $b$ to zero?*

### 18.1.1 Solving for the energy-optimal control

As usual, we construct the Hamiltonian

$$H(\lambda, \mathbf{x}, u) = |u| - \lambda_1 x_2 - \lambda_2 u$$

and use it to generate the costate equations

$$\dot{\lambda}_1 = \frac{\partial H}{\partial x_1} = 0 \quad \text{and} \quad \dot{\lambda}_2 = \frac{\partial H}{\partial x_2} = -\lambda_1,$$

which lead to the usual costates

$$\lambda_1 = c_1 \quad \text{and} \quad \lambda_2 = c_2 - c_1 t.$$

**Problem 18.1.** *Use Pontryagin's principle to express the possible energy-optimal controls (nonsingular and singular) in terms of the costate $\lambda_2$ given that $|u(t)| \leq 1$.*

https://doi.org/10.1515/9783111290157-018

**Nonsingular optimal control patterns**

To determine what sort of nonsingular optimal control switching is possible, we can consider the schematic shown in Figure 18.1, where the horizontal regions correspond to the optimal controls as follows:

$$1 < \lambda_2 \quad \Longleftrightarrow \quad u^* = 1,$$
$$-1 < \lambda_2 < 1 \quad \Longleftrightarrow \quad u^* = 0,$$
$$\lambda_2 < -1 \quad \Longleftrightarrow \quad u^* = -1.$$

The seven numbered lines represent the various possible versions of the costate function $\lambda_2(t) = c_2 - c_1 t$ that generate different nonsingular optimal control switches in the following patterns:

line 1 $\quad\Longrightarrow\quad u^* = \boxed{-1 \to 0 \to 1}$ or $\boxed{-1 \to 0}$ or $\boxed{-1}$ (depending on $T^*$)

line 2 $\quad\Longrightarrow\quad u^* = \boxed{0 \to 1}$ or $\boxed{0}$ (depending on $T^*$)

line 3 $\quad\Longrightarrow\quad u^* = \boxed{1}$

line 4 $\quad\Longrightarrow\quad u^* = 0$

line 5 $\quad\Longrightarrow\quad u^* = -1$

line 6 $\quad\Longrightarrow\quad u^* = \boxed{1 \to 0 \to -1}$ or $\boxed{1 \to 0}$ or $1$ (depending on $T^*$)

line 7 $\quad\Longrightarrow\quad u^* = \boxed{0 \to -1}$ or $0$ (depending on $T^*$)

Altogether, there are nine distinct nonsingular optimal control switching patterns (repeats are not boxed), and we note that an optimal control must spend more than an instant of time at 0 when switching between 1 and −1 (or vice versa).

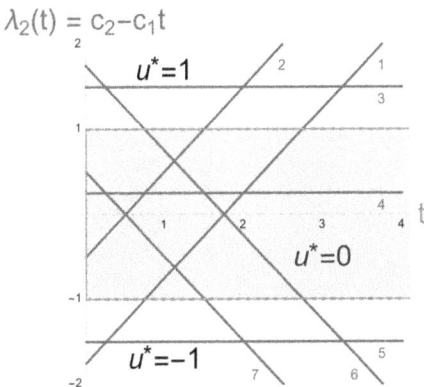

**Figure 18.1:** Costate possibilities for energy minimization.

**Lab Activity 8.**

1.  *Sketch the energy-optimal switching curve and describe the optimal control from each $(a, b)$ to $(0, 0)$ in the $x_1 x_2$-plane for the nanosatellite controlled with $|u(t)| \leq 1$.*
2.  *Sketch the region of controllability for the nanosatellite (with bounded acceleration $|u(t)| \leq 1$) being optimally controlled to $(0, 0)$ at $T^* = 5$ seconds while minimizing energy consumed.*

# 19 Time & energy optimal control

We have already determined how to control the nanosatellite to $(0,0)$ in the $x_1 x_2$-plane time-optimally and energy-optimally. In this chapter, we determine how to control the nanosatellite to $(0,0)$ while minimizing the time and energy simultaneously.

## 19.1 Review

In Lab Activity 4, we determined the time-optimal control synthesis shown in Figure 19.1. For instance, to move from $(-5,1)$ to $(0,0)$ time-optimally, we follow a $u^* = 1$ trajectory up to the $u^* = -1$ part of the switching curve and then follow that part of the switching curve to $(0,0)$ as in Figure 19.1. In Lab Activity 8, we determined the energy-optimal control synthesis shown in Figure 19.2. To move from $(-5,1)$ to $(0,0)$ energy-optimally, we could choose any path that never increases the $x_2$-coordinate as it moves toward the $u^* = -1$ part of the switching curve. One example follows the $u^* = 0$ (horizontal) trajectory to the right as in Figure 19.2. Since the velocity (measured by $x_2$) is slower nearer the $x_1$-axis, these trajectories take more time than ones farther from the $x_1$-axis (like the $u^* = 1$ piece of the time-optimal trajectory from $(-5,1)$ in Figure 19.1) but use less energy (since they do not change $x_2$ as much overall).

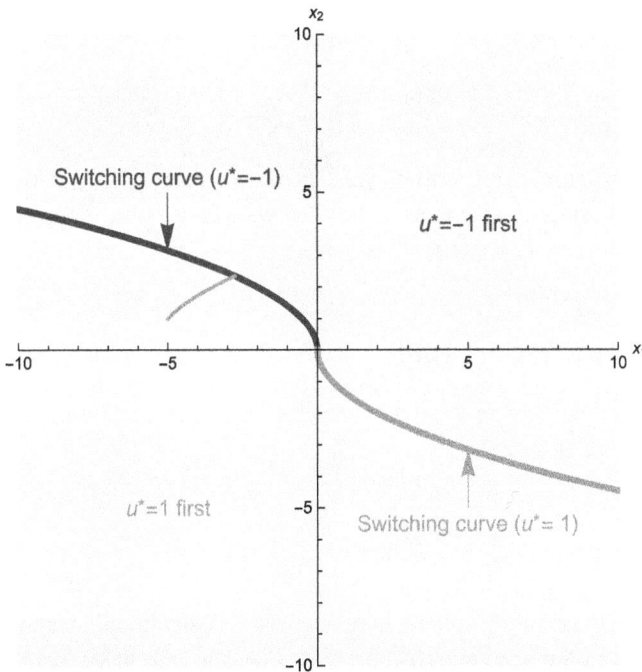

**Figure 19.1:** Time-optimal control synthesis.

https://doi.org/10.1515/9783111290157-019

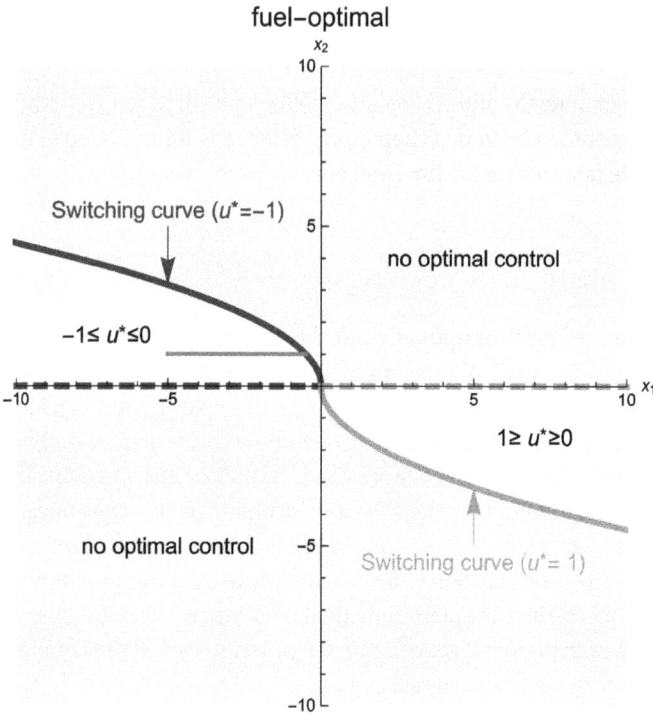

**Figure 19.2:** Time-optimal control synthesis.

**Question 19.1.** *Where are the initial $(a, b)$ from which the time-optimal trajectory happens to also be energy optimal?*

In summary, the time-optimal trajectories typically use more energy, and the energy-optimal trajectories typically take more time. Now we will see what happens if we try to minimize the time and energy simultaneously.

## 19.2  Time & energy optimal control

To optimize with respect to time and energy simultaneously, it is useful to minimize the cost functional

$$\int_0^T \tau + |u|\, dt,$$

where $\tau > 0$ weighs the importance of time relative to energy consumption (adapted from [2, Example 5.12]). The higher the value of $\tau$, the more important the time.

**Problem 19.1.** *Show that singular time & energy optimal controls are not possible.*

We conclude that the time & energy optimal controls are of the form

$$u^* = \begin{cases} 0 & \text{if } |\lambda_2(t)| < 1, \\ \text{sgn}(\lambda_2) & \text{if } |\lambda_2(t)| > 1, \end{cases}$$

so that time & energy optimal control switching is dictated (as in Problem 18.1 for non-singular energy-optimal control) by the schematic in Figure 19.3, where the numbered lines represent different options for the costate function $\lambda_2(t) = c_2 - c_1 t$. As a result, there are nine different possible time & energy optimal control switching patterns:

$$u^* = -1 \rightarrow 0 \rightarrow 1$$
$$u^* = -1 \rightarrow 0$$
$$u^* = 0 \rightarrow 1$$
$$u^* = 1$$
$$u^* = 0$$
$$u^* = -1$$
$$u^* = 1 \rightarrow 0 \rightarrow -1$$
$$u^* = -1 \rightarrow 0$$
$$u^* = 0 \rightarrow -1,$$

where we have ruled out the three cases ending in control $u^* = 0$ since those fix the $x_2$-level, which would only work if we had already arrived at the target $(0, 0)$.

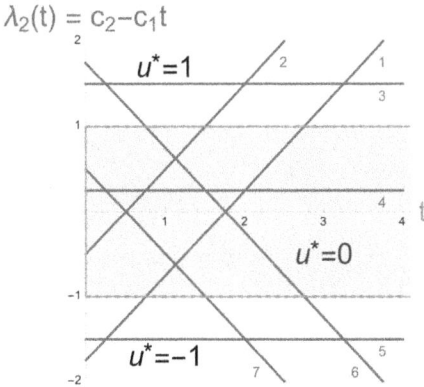

**Figure 19.3:** Costate possibilities for time & energy minimization.

Notice that before an optimal control switches from 1 to −1, the costate $\lambda_2(t) = c_2 - c_1 t$ spends some amount of time, $\Delta t$, in the region while $u^* = 0$. During this time, $\lambda_2$ covers $\Delta \lambda_2 = -2$ vertical units since it moves from $\lambda_2 = 1$ down to $\lambda_2 = -1$. This means that the slope $-c_1$ of $\lambda_2(t) = c_2 - c_1 t$ is equal to $\frac{-2}{\Delta t}$. We can then solve the resulting equation $-c_1 = \frac{-2}{\Delta t}$ for $\Delta t = \frac{2}{c_1}$. A similar argument shows that $\Delta t = \frac{2}{-c_1}$ between an optimal control switch from $-1$ to $1$ (in which case its slope $-c_1$ is positive; e. g., see line 1 in Figure 19.3). As a result, we can say that in either case the optimal control must stay at 0 for

$$\Delta t = \frac{2}{|c_1|} \tag{19.1}$$

time units between switches from 1 to $-1$ (or vice versa).

## 19.3 Time & energy optimal switching curve

To construct the switching curve for the time & energy optimal control, we start as usual with the trajectories associated with each optimal control value that lead directly to the target $(0, 0)$.

**Question 19.2.** *Given that the state equations are $\dot{x}_1 = x_2$ and $\dot{x}_2 = u$, what does the trajectory to $(0, 0)$ associated with $u^* = 0$ look like in the $x_1 x_2$-plane?*

For the optimal controls $u^* = 1$ and $u^* = -1$, the trajectories to $(0, 0)$ are shown in Figure 19.4 and are the same as in the time-optimal or energy-optimal cases (since the state equations are unchanged).

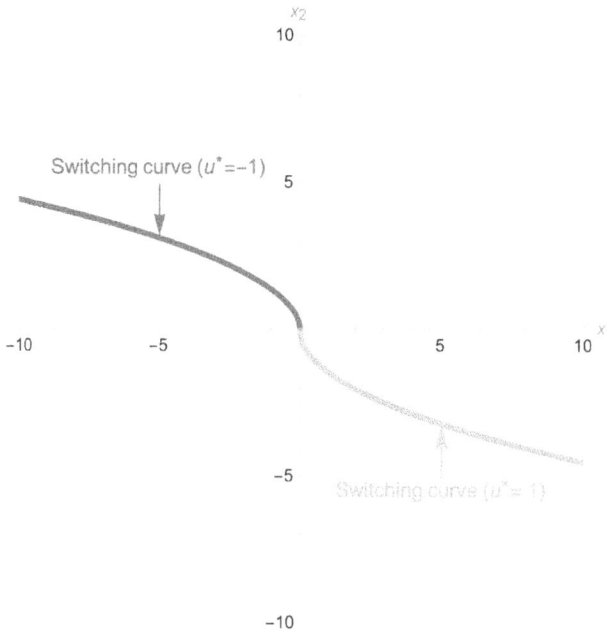

**Figure 19.4:** Time & energy optimal trajectories to $(0, 0)$.

We have already seen in (19.1) that the control must switch to 0 for $\frac{2}{|c_1|}$ time units between a switch from 1 to –1 (or vice versa), so the $u^* = 0$ part of the switching curve could be constructed by tracking $\frac{2}{|c_1|}$ time units back from each of the $u^* = 1$ and $u^* = -1$ parts of the switching curve. However, since the value of $c_1$ is a mystery, we need to take a more general approach to finding the $u^* = 0$ part of the switching curve, which relies on the following fact.

### 19.3.1 The Hamiltonian is constant

It turns out that the Hamiltonian is always constant along the optimal inputs:

$$H(\lambda^*(t), \mathbf{x}^*(t), \mathbf{u}^*(t)) = \text{constant} \quad \text{for all } t.$$

The single-state version of this follows immediately from the (simplified) integral Euler–Lagrange equation

$$f - \dot{x} f_{\dot{x}} = \text{constant}$$

applied to the integrand $\lambda \dot{x} + H(\lambda, x, u)$:

$$\lambda^*(t)\dot{x}(t) + H(\lambda^*(t), x^*(t), u^*(t)) - \dot{x}(t)\lambda^*(t) = \text{constant}$$

$$\Downarrow$$

$$H(\lambda^*(t), x^*(t), u^*(t)) = \text{constant}.$$

We could have introduced this result earlier, but we have saved it until now since this is the first time we make use of it. Note that when there is a free end time $T$, we know that the Hamiltonian is actually constant at zero along the optimal inputs since $H(T^*) = 0$.

### 19.3.2 The $u^* = 0$ part of the switching curve

Since there is a free end time in the nanosatellite problem, we conclude that the optimal control $u^* = 0$ satisfies the equation

$$0 = H(\lambda^*(t), \mathbf{x}^*(t), u^*(t))$$
$$= \tau + |u^*(t)| - \lambda_1^*(t)x_2^*(t) - \lambda_2^*(t)u^*(t)$$
$$= \tau - c_1 x_2^*(t).$$

Solving this for $x_2^*(t)$ gives us the constant vertical component $x_2^*(t) = \frac{\tau}{c_1}$ in the $x_1 x_2$-plane of the entire (horizontal) trajectory associated with $u^* = 0$. In Figure 19.5, we have labeled this constant vertical component as $b_0 = \frac{\tau}{c_1}$, whereas $a_0$ is the $x_1$-coordinate at which the switch from $u^* = 1$ to $u^* = 0$ is made. Notice that $b_0$ in this case is always positive.

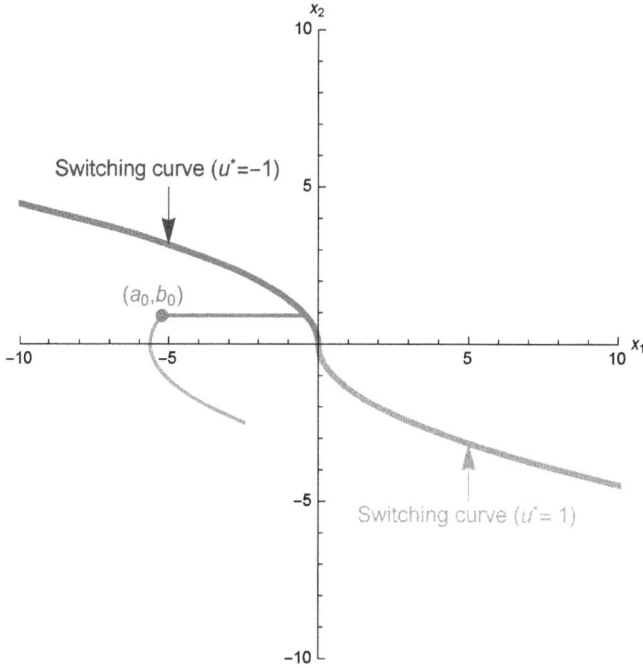

**Figure 19.5:** Time & energy optimal trajectory using $u^* = 0$.

Solving the state equations $\dot{x}_1 = x_2$ and $\dot{x}_2 = 0$ with $u^* = 0$ and initial state $(a_0, b_0)$, we get

$$x_1^*(t) = a_0 + b_0 t \quad \text{and} \quad x_2^*(t) = b_0,$$

and we already saw in (19.1) that this piece of the optimal trajectory lasts for $\frac{2}{|c_1|}$ time units. We can translate this via $b_0 = \frac{\tau}{c_1}$ (by solving for $c_1 = \frac{\tau}{b_0}$) into the more useful $\frac{2|b_0|}{\tau} = \frac{2b_0}{\tau}$ time units (since $b_0$ in this case is positive), which no longer contains the mystery constant $c_1$. We conclude that the terminal point of the $u^* = 0$ (horizontal) trajectory, where it meets the $u^* = -1$ part of the switching curve, is located at

$$\left(a_0 + b_0 \frac{2b_0}{\tau}, b_0\right). \tag{19.2}$$

Using this point (19.2) as the initial state and solving the state equations $\dot{x}_1 = x_2$ and $\dot{x}_2 = -1$ (since $u^* = -1$), we get the parameterization

$$x_1^*(t) = a_0 + b_0 \frac{2b_0}{\tau} + b_0 t - \frac{t^2}{2} \quad \text{and} \quad x_2^*(t) = b_0 - t$$

of the $u^* = -1$ part of the switching curve from the initial state given in (19.2) to the target state $(0, 0)$. Evidently, this hits $(0, 0)$ at $t = b_0$ (when $x_2^*(t) = 0$), so we conclude that

$$0 = x_1^*(b_0) = a_0 + b_0 \frac{2b_0}{\tau} + b_0 b_0 - \frac{b_0^2}{2}$$

$$\Downarrow$$

$$a_0 = -b_0 \frac{2b_0}{\tau} - \frac{b_0^2}{2} = -\left(\frac{1}{2} + \frac{2}{\tau}\right) b_0^2.$$

Since $(a_0, b_0)$ is the state in the $x_1 x_2$-plane where the $u^* = 0$ trajectory begins, the set of all such points gives the $u^* = 0$ part of the switching curve. Thus we have derived the formula

$$x_1 = -\left(\frac{1}{2} + \frac{2}{\tau}\right) x_2^2$$

for the $u^* = 0$ part of the switching curve on the $x_2 \geq 0$ half of the $x_1 x_2$-plane. A similar argument gives the formula

$$x_1 = \left(\frac{1}{2} + \frac{2}{\tau}\right) x_2^2 \tag{19.3}$$

for the $u^* = 0$ part of the switching curve on the $x_2 < 0$ half of the $x_1 x_2$-plane. Figure 19.6 gives the time & energy optimal control synthesis for the case where $\tau = 1$. Notice that

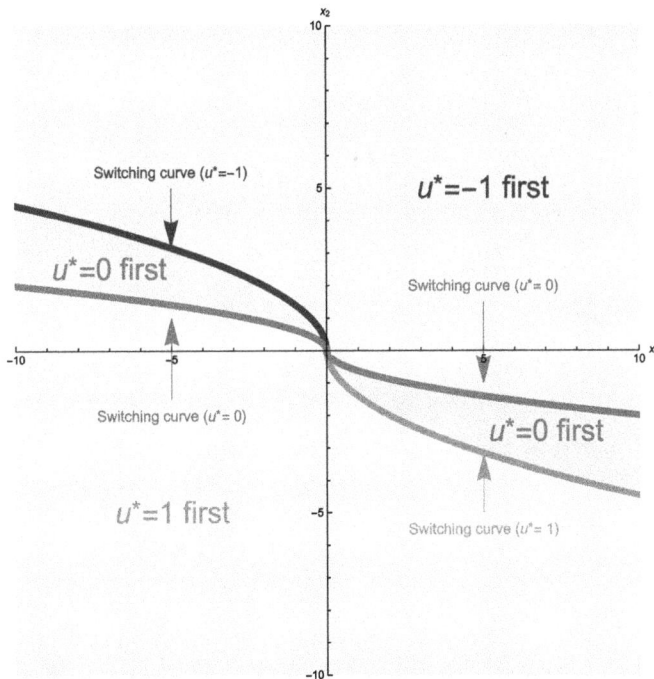

**Figure 19.6:** Time & energy optimal synthesis.

the possible optimal control switching patterns

$$u^* = -1 \rightarrow 0 \rightarrow 1$$
$$u^* = 0 \rightarrow 1$$
$$u^* = 1$$
$$u^* = -1$$
$$u^* = 1 \rightarrow 0 \rightarrow -1$$
$$u^* = 0 \rightarrow -1$$

allow us to deduce that we should always start with $u^* = 0$ for initial states in the regions between the $u^* = 0$ part of the switching curve and the $u^* = -1$ or $u^* = 1$ parts of the switching curve.

**Problem 19.2.**

a) *Give the argument establishing formula (19.3) for the $u^* = 0$ part of the switching curve on the $x_2 < 0$ half of the $x_1x_2$-plane.*

b) *Explain what happens to the $u^* = 0$ part of the switching curve as $\tau \rightarrow \infty$.*

c) *Explain what happens to the $u^* = 0$ part of the switching curve as $\tau \rightarrow 0$.*

**Lab Activity 9.** *Give the total time needed to time & energy-optimally control the nano-satellite from $(-1, -1)$ to $(0, 0)$ (using weight $\tau = 1$).*

# A  Invitation to research

## Description

Explore journal papers from the past decade in a field of interest to you and find a continuous-time optimal control problem that you would like study. Read the paper(s) carefully and thoroughly until you are comfortable in your understanding of how the ideas from this text intersect with the optimal control problem you chose. Write an abstract, a poster, and a short paper with a description of the problem, the solution method, and the solution, and provide an appropriate citation of your source(s).

## Research tips

To begin, you can try searching "optimal control" paired with a keyword or two that identify an area of interest, for example, "optimal control medical", or, more narrowly, "optimal control chemotherapy". Google is probably fine, but you can also try Google Scholar for a search that should return only the kinds of materials you will want (e. g., you can choose a particular time range). You can extend your search by locating papers cited in the bibliographies of the first sources you find.

## Abstract

An abstract is a paragraph outlining the problem (you should use prose and as little notation as possible) and should include appropriate citation of your source(s). You may use any style of citation you like, but it should include all of the standard information that will allow someone to find the source(s) in a library or database. If you have a link for a source, then you should include that.

## Poster

Your poster should use exactly the notation ($x_i$ for states and $u_i$ for controls) from this text and should identify:
- title of your problem
- citation of your source
- states $x_i$
- controls $u_i$
- state equations
- cost functional and/or end-state cost to be minimized
- solution

You can use the sample at the end of this chapter as a template.

https://doi.org/10.1515/9783111290157-020

**Paper**

Your paper should begin with your abstract and continue with a more thorough summary of the problem, including at least the following elements:
- A description of what makes this problem interesting.
- A clear identification of the variable(s) involved, the cost functional and/or end-state cost, and any constraints. You should use the same notation as in this text for optimal control problems, even if your source uses a different notation. For example, use $x_i$ for states and $u_i$ for controls, and if the problem you study uses maximization, then make sure to negate the cost to convert it for minimization.
- A description of the solution process and the optimal solutions identified.
  Make sure to use the same conventions from this text for your description—for example, when constructing the Hamiltonian, identifying the costate equations, and applying Pontryagin's principle to minimize the Hamiltonian—and make sure to note where your paper uses different conventions.
  If your paper uses numerical methods to approximate solutions to differential equations rather than solving for explicit solutions, then note that and indicate the methods used.

## Poster template

**The Application of Optimal Control Theory to the General Life History Problem**

From W. Schaffer, in *The American Naturalist*, 1983, Volume 121, pages 418–431

### States

$x_{1,i}(t)$ = number of leaves in season $i$

$x_{2,i}(t)$ = amount of stored energy in season $i$

$x_{3,i}(t)$ = number of seeds in season $i$

### Controls

$u_{1,i}(t)$ = fraction of fixed carbon devoted to leaves in season $i$

$u_{2,i}(t)$ = fraction of fixed carbon stored in season $i$

$u_{3,i}(t)$ = fraction of fixed carbon devoted to seeds in season $i$

### State equations

$$\dot{x}_{1,i} = \frac{r}{\text{cost}_1} x_{1,i} u_{1,i}, \quad \dot{x}_{2,i} = \frac{r}{\text{cost}_2} x_{1,i} u_{2,i}, \quad \dot{x}_{3,i} = \frac{r}{\text{cost}_3} x_{1,i} u_{3,i}$$

($r$ = photosynthesis rate)

### Cost functional

$$\int_0^{t_1} -\dot{x}_3 \, dt = -\text{total number of seeds in lifetime of length } t_1$$

($x_3(t) = x_{3,i}(t)$ for $t$ in season $i$)

### Solution

Pontryagin's principle shows that in each season, the plant first makes leaves, then either stores the rest of its energy or produces seeds (but not both). This accurately describes the strategy of some perennial plants that lose their leaves at the end of each season.

# B Solutions

## Answers to questions

**Answer 2.1:** We know that $x \in C^2$ on $[t_0, t_1]$ because $x^*$ and $\eta$ are both $C^2$ on $[t_0, t_1]$. Since $\eta(t_0) = \eta(t_1) = 0$, we know that $x(t_0) = x^*(t_0) = x_0$ and $x(t_1) = x^*(t_1) = x_1$.

**Answer 2.2:** It must equal zero for all $t \in [t_0, t_1]$.

**Answer 3.1:** There is no solution, since we can increase $T$ to get more and more area.

**Answer 3.2:** We have so far ignored the constraint that the rope has a fixed length (which we could encode via $\int_0^T \sqrt{1 + \dot{x}^2}\, dt = $ constant).

**Answer 3.3:** A vertical line in the $tx$-plane at $t = t_1$.

**Answer 3.4:** In this case the target curve is represented by a constant function $\mathrm{tar}(t) = x_1$ (with zero derivative), so the transversality condition reduces to $f(T^*) = \dot{x}^*(T^*) f_{\dot{x}}(T^*)$.

**Answer 3.5:** In this case the target curve is a vertical line, so $\mathrm{tar}'(T^*) = \infty$. Dividing the transversality condition by $\mathrm{tar}'(T^*)$ thus results in $f_{\dot{x}}(T^*) = 0$.

Notice that this case corresponds to $T = T^*$ in our derivation of the transversality condition, so that argument would need to be modified to carefully justify the special case of transversality $f_{\dot{x}}(T^*) = 0$ in this case.

**Answer 4.1:** Anything continuous whose graph connects the endpoints via a finite number of straight-line pieces having slope $\pm 1$ as in Figure B.1.

**Answer 4.2:** The integrand is always nonnegative, so zero is the lowest value the integral functional can attain, and these $\mathcal{D}^1$ functions all return zero in the integral functional.

**Answer 4.3:** Smooth out the sharp corners as in Figure B.2.

**Answer 4.4:** No such minimizer exists, since we can always reduce the integral value toward zero by smoothing the corners more conservatively, but we can only get to a zero integral value with the $\mathcal{D}^1$ function. Note also that the "smoothed" functions do not satisfy the Euler–Lagrange equation (which a minimizer must) since the smoothed corners correspond to nonzero second derivatives.

**Answer 6.1:** Any point on the $x$-axis other than the origin.

**Answer 6.2:** The origin (which happens to be covered by every function).

https://doi.org/10.1515/9783111290157-021

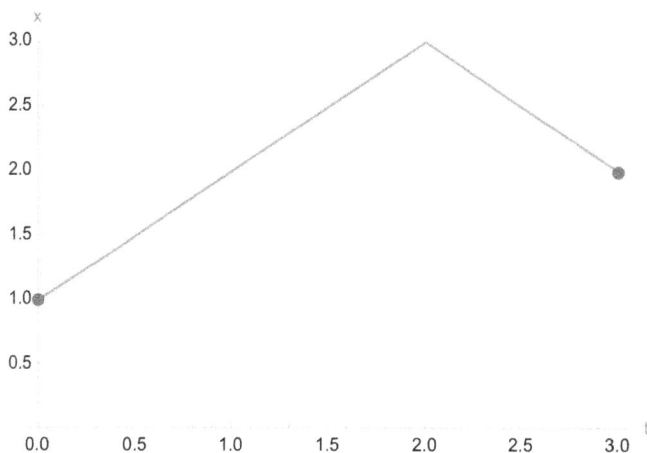

**Figure B.1:** A $\mathcal{D}^1$ solution.

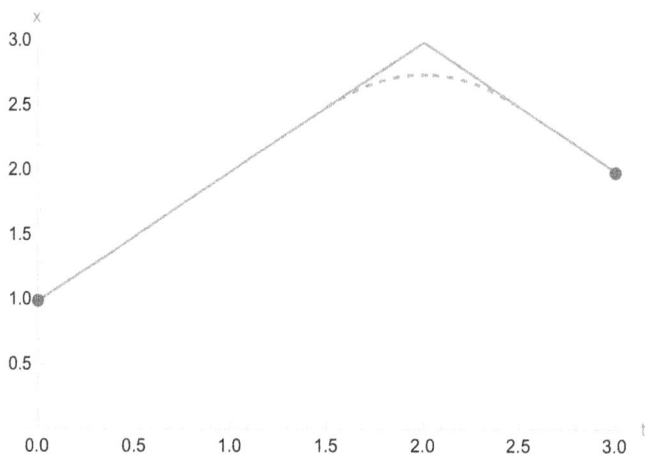

**Figure B.2:** Smoothed $\mathcal{D}^1$ function.

**Answer 6.3:** $p(t, x) = 1$ and $p(t, x) = a = \frac{x}{t}$.

**Answer 6.4:** $K[x] = J[x]$ for any function $x(t)$ in the field of extremals since $p(t, x(t)) = \dot{x}(t)$ in that case. Since $x^*$ is always in the field of extremals, we know that $K[x^*] = J[x^*]$.

**Answer 8.1:** $x_0(t + 1) = \sum_{i=12}^{49} a_{0,i} x_i(t)$ since only elephants between the ages of 12 and 49 can produce calves.

**Answer 8.2:** The new 10–19 year group includes some former 10–19 year-olds (i. e., the 10–18 year-olds who survived the year) in addition to any 9 year-olds who survived the year.

**Answer 8.3:** $\lambda \geq 1$.

**Answer 8.4:** The expected number of breeding offspring from an individual in the $i$th age class.

**Answer 9.1:** $\infty$: we have one equation for each time $t$ in the continuum of the interval $[t_0, T]$.

**Answer 11.1:** Let $x_1(t) = x(t)$ be the position. Introduce a "velocity" state $x_2(t)$: $\dot{x}_1 = x_2$ and $\dot{x}_2 = u$.

**Answer 15.1:** The entire $x_1 x_2$-plane.

**Answer 15.2:** The entire $x_1 x_2$-plane.

**Answer 15.3:** Not sure, but perhaps not the entire $x_1 x_2$-plane.

**Answer 16.1:** The $x_1$-axis in the $x_1 x_2$-plane.

**Answer 16.2:** The costate vector $\lambda_1(T^*)\vec{i} + \lambda_2(T^*)\vec{j}$ at $T^*$ is vertical: $\lambda_1(T^*) = 0$.

**Answer 18.1:** The total acceleration used (left or right).

Note that this assumes that energy is proportional to total acceleration and then simply uses a proportion factor of 1 since the energy-minimizer is the same regardless of the proportion factor.

**Answer 18.2:** The total $x_2$ (vertical) distance covered by the trajectory between times 0 and $T$ since

$$\int_0^T |u|\, dt = \int_0^T |\dot{x}_2|\, dt$$

$$= \text{the total (unsigned) change in } x_2(t) \text{ from 0 to } T.$$

**Answer 18.3:** $|b|$.

**Answer 19.1:** Along the time-optimal switching curve since those trajectories move directly to the target at $(0,0)$ without changing the $x_2$-coordinate more than necessary.

**Answer 19.2:** The single point $(0,0)$. Since $\dot{x}_2 = 0$ means no change in the $x_2$-coordinate, and we would need $x_2 = 0$ to pass through $(0,0)$. Then $\dot{x}_1 = 0$ means no change in the $x_1$-coordinate, and we would need $x_1 = 0$ to pass through $(0,0)$.

## Solutions to problems

### Problem 1.1 (solution)

(a) These functions generate the output values

$$J[x_\epsilon] = \int\limits_0^1 \sqrt{\frac{1 + \dot{x}_\epsilon^2}{1 - x_\epsilon}}\, dt = \int\limits_0^1 t^{-\frac{\epsilon}{2}} \sqrt{1 + \epsilon^2 t^{2\epsilon - 2}}\, dt.$$

(b) Plugging $\epsilon = 1$ into the integral formula gives

$$J[x_1] = \int\limits_0^1 t^{-\frac{1}{2}} \sqrt{2}\, dt = 2\sqrt{2}.$$

The graph of the function $x_1(t) = 1 - t$ is a straight line through the points $(0, 1)$ and $(1, 0)$.

### Problem 1.2 (solution)

(a) If $\dot{x}(t)$ is continuous ($x \in C^1$), it is certainly piecewise continuous ($x \in \mathcal{D}^1$). It follows that we have the relationships

$$C^2 \subseteq C^1 \subseteq \mathcal{D}^1 \subseteq C^0.$$

This means that we have restricted the original brachistochrone problem somewhat (since its points only had to follow a continuous path from $A$ to $B$). It will turn out that we could safely restrict the problem even further to $x \in C^2$, since the solution to the original brachistochrone problem happens to satisfy that stronger property.

(b) Anything continuous works as long as it has a finite number of sharp corners. For example, the graph of the absolute value function $f(x) = |x|$ has a sharp corner at the origin, as shown in Figure B.3. The derivative of the absolute value function is

$$f'(x) = \begin{cases} -1 & \text{for } x < 0, \\ ? & \text{for } x = 0, \\ 1 & \text{for } x > 0, \end{cases}$$

which is discontinuous only at the origin $x = 0$. Notice that there is no unique tangent line to the graph of $f$ at $(0, 0)$, whose slope we could otherwise assign as the value of $f'(0)$. Another convention is to assign the entire interval $[-1, 1]$ since there are tangent lines to the graph of $f$ at $(0, 0)$ having all of those slopes. In this case though, the notation changes from the derivative (which does not exist there) to a "generalized" derivative.

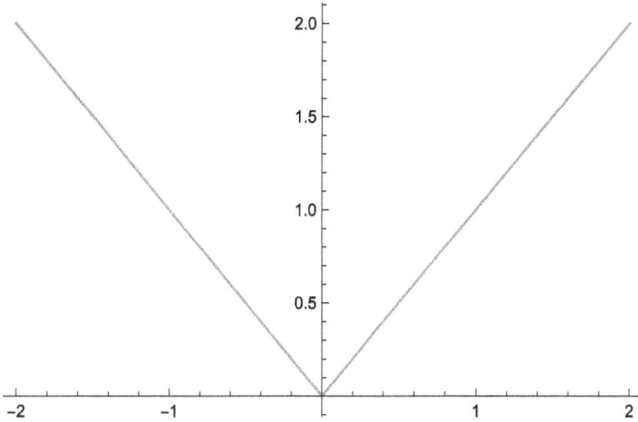

**Figure B.3:** Graph of absolute value function.

## Problem 2.1 (solution)

In this case, we have $f(t, x, \dot{x}) = (\dot{x}t)^2$, so the Euler–Lagrange equation translates into

$$0 - \frac{d}{dt}(2\dot{x}t^2) = 0. \tag{B.1}$$

Computing the derivative with respect to $t$ results (via the product rule) in the second-order differential equation

$$-2(\ddot{x}t^2 + 2t\dot{x}) = 0,$$

which we can (algebraically) solve for $\ddot{x}$ to get

$$\ddot{x} = \frac{-2\dot{x}}{t}. \tag{B.2}$$

This can be solved (e. g., using Python or Mathematica) to give

$$x(t) = -\frac{c_1}{t} + c_2 \tag{B.3}$$

in terms of two arbitrary constants $c_1$ and $c_2$.

If you want to find extremals without resorting to software, there is a trick that avoids a second-order differential equation altogether and that works whenever $f_x = 0$. We simply notice that (what is left of) the Euler–Lagrange equation $-\frac{d}{dt}(f_{\dot{x}}(t, x^*, \dot{x}^*)) = 0$ can be used, without computing the derivative with respect to $t$, to conclude that $f_{\dot{x}}(t, x^*, \dot{x}^*)$ is constant with respect to $t$. For example, in the present problem, this results in

$$\dot{x}t^2 = c_1 \quad \Rightarrow \quad \dot{x} = \frac{c_1}{t^2} \quad \Rightarrow \quad x(t) = -\frac{c_1}{t} + c_2,$$

which only required simple integration in the final step to obtain the same solution (B.3) as above.

In any case, to satisfy the endpoint conditions $x(1) = 0$ and $x(2) = 1$, we choose $c_1 = 2$ and $c_2 = 2$:

$$x^*(t) = 2 - \frac{2}{t}.$$

(Notice that this function is $C^2$ on the interval $[1, 2]$ but not at $t = 0$. If the interval of integration had instead included 0, then we would have had to address that issue.)

### Problem 3.1 (solution)

1.  The Euler–Lagrange equation in this case reduces to

$$\ddot{x} = \frac{-2\dot{x}}{t} \quad \text{or} \quad \dot{x}t^2 = \text{constant}$$

with solution $x^*(t) = -\frac{c_1}{t} + c_2$.

2.  Applying the endpoint condition $x(1) = 0$ and the hits-the-target condition gives

$$x^*(t) = \frac{t-1}{t(T-1)T}.$$

3.  The transversality condition (with $f(t, x, \dot{x}) = (\dot{x}t)^2$ and $\text{tar}(t) = \frac{1}{t^2}$) gives us that $T^* = \frac{3}{4}$.

An extremal is thus $x^*(t) = \frac{16}{3t} - \frac{16}{3}$.

### Problem 4.1 (solution)

The Euler–Lagrange equation in this case gives

$$\frac{d}{dt}(2(\dot{x}^2 - 1)2\dot{x}) = 0,$$

which implies that

$$4\dot{x}(\dot{x}^2 - 1) \text{ is constant} \quad \Longrightarrow \quad \dot{x} \text{ is constant},$$

so the extremal $x^*$ is the straight line connecting the endpoints, $x^*(t) = \frac{1}{3}t + 1$. Direct comparison then gives

$$J[x] - J[x^*] = \int_0^3 \left(\left(\frac{1}{3} + \dot{\eta}\right)^2 - 1\right)^2 - \left(\left(\frac{1}{3}\right)^2 - 1\right)^2 dt$$

$$= \int_0^3 \left( -\frac{32\dot{\eta}}{27} - \frac{4\dot{\eta}^2}{3} + \frac{4\dot{\eta}^3}{3} + \dot{\eta}^4 \right) dt,$$ (B.4)

which is not obviously nonnegative. Indeed, Figure B.4 shows that the integrand is some-times positive and sometimes negative depending on the value of $\dot{\eta}$.

Integrand as a function of $\dot{\eta}$

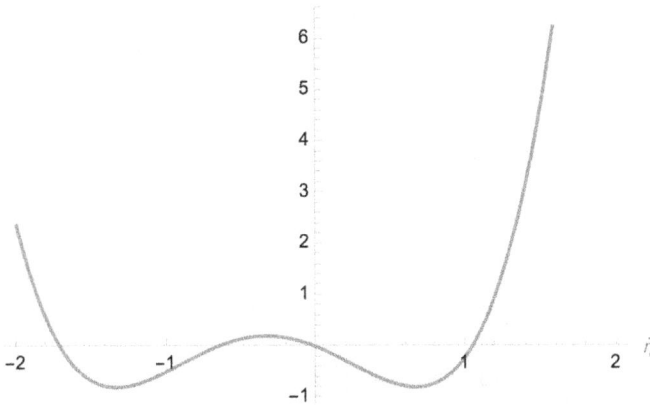

**Figure B.4:** The integrand is sometimes positive and sometimes negative.

We certainly cannot conclude that $x^*$ is a minimizer, but we might be able to use this image to guide the construction of an admissible $x$ whose corresponding $\dot{\eta} = \dot{x} - \dot{x}^* = \dot{x} - \frac{1}{3}$ stays in the "negative" zone (in which case, we could conclude that $x^*$ is certainly not a minimizer). For instance, if we can construct an admissible $x(t)$ for which $\dot{x}(t) \in [1/3, 4/3]$ for all $t \in [0, 3]$, then we know that the corresponding $\eta = x - x^*$ has $\dot{\eta}(t) \in [0, 1]$ for all $t \in [0, 3]$. According to Figure B.4, this would ensure that the integrand in (B.4) is always negative. Then equation (B.4) would allow us to conclude that $J[x] - J[x^*] < 0$, which would mean that $x^*$ is certainly not a minimizer.

**Problem 4.2 (solution)**
We can compute the integral functional value $J[x^*] = 24$, and if there was a function $x$ in $\mathcal{D}^1$ with a strictly lower integral value $J[x]$ (say, 22), then we could smooth the corners on the graph of $x$ in such a way as to get arbitrarily close to that lower value with a $C^2$ function. This would mean that there is a $C^2$ function with strictly lower integral value than our $C^2$ minimizer, which is not possible.

**Problem 5.1 (solution)**

(a) $f_{\ddot{x}\ddot{x}} = 2t^3$ is nonnegative on the interval of integration $[1, 2]$, regardless of the extremal $x^*$, so the most we can conclude from the second-derivative test is that the extremal $x^* = 4 - \frac{4}{t^2}$ *might* be a minimizer.

(b) $f_{\dot{x}\dot{x}} = 12\dot{x}^2 - 4$, which evaluates to $-\frac{8}{3}$ for the extremal $x^* = \frac{t}{3} + 1$ of this problem, so the second-derivative test implies that the extremal $x^* = \frac{t}{3} + 1$ is *certainly not* a minimizer. Therefore there cannot be any minimizer for this problem (since any minimizer is an extremal, and the only extremal is not a minimizer).

(c) $f_{\dot{x}\dot{x}} = \frac{2}{\dot{x}^3}$, which evaluates to 2 for the extremal $x^* = t$ of this problem, so the most we can conclude from the second-derivative test is that the extremal $x^* = t$ *might* be a minimizer.

**Moral:** The second-derivative test is relatively simple to check but can only be used to rule out a minimizer.

**Problem 5.2 (solution)**

(a) There are many different functions that work, but one is defined by

$$x(t) = \begin{cases} 3t & \text{if } t \in [0, 0.5], \\ -t + 2 & \text{if } t \in [0.5, 1], \end{cases}$$

and graphed in Figure B.5. This has the integral value $J[x] = -\frac{1}{3}$, which is clearly less than the integral value $J[x^*] = 1$ of the extremal $x^* = t$.

Notice that this means that the extremal $x^* = t$ cannot be a minimizer, even in $\mathcal{C}^2$, because if it were, then it would also be a minimizer in $\mathcal{D}^1$.

(b) $V_1 = 0$ in this case since the extremal satisfies the Euler–Lagrange equation. We can also see this directly since $f_x = 0, f_{\dot{x}} = -\frac{1}{(\dot{x}^*)^2} = -1$, and

$$\eta = \frac{x - x^*}{\epsilon} = \begin{cases} \frac{2t}{\epsilon} & \text{if } t \in [0, 0.5], \\ \frac{2 - 2t}{\epsilon} & \text{if } t \in [0.5, 1], \end{cases}$$

which leads to

$$V_1 = \int_0^{0.5} -\frac{2}{\epsilon} \, dt + \int_{0.5}^1 \frac{2}{\epsilon} \, dt = 0.$$

We compute $V_2$ using $f_{xx} = 0, f_{x\dot{x}} = 0$, and $f_{\dot{x}\dot{x}} = \frac{2}{(\dot{x}^*)^3} = 2$:

$$V_2 = \frac{1}{2}\left( \int_0^{0.5} \left(\frac{2}{\epsilon}\right)^2 2 \, dt + \int_{0.5}^1 \left(-\frac{2}{\epsilon}\right)^2 2 \, dt \right) = \frac{4}{\epsilon^2}.$$

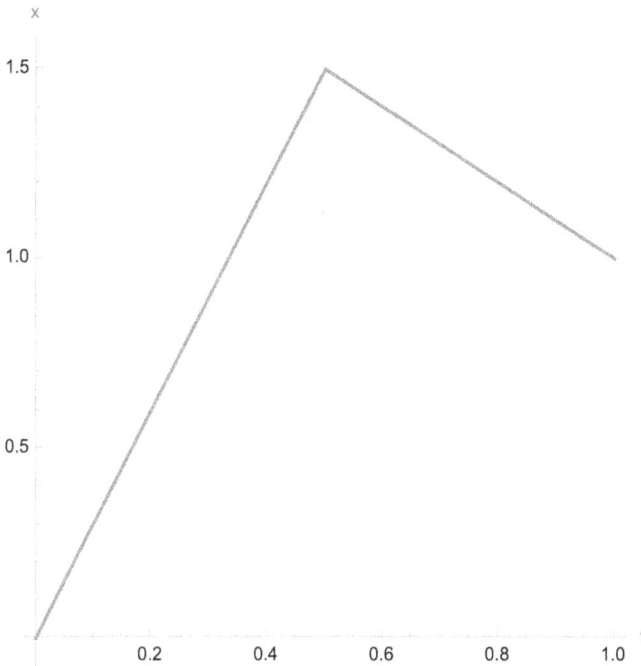

**Figure B.5:** Graph of $x(t)$.

Since $V_1 = 0$ and $V_2$ is positive, the $O(\epsilon^3)$ term in $\Delta J$ must be negative and larger in magnitude than $\epsilon^2 V_2 = 4$. Note that the "variation" $x$ here is fixed, so that the value of $\epsilon$ is also essentially fixed in this construction. The formula $\eta = \frac{x-x^*}{\epsilon}$ makes it look like $\eta$ varies with $\epsilon$; however, the original version of this formula $x = x^* + \epsilon \eta$ highlights that it is the variations $x$ that vary with $\epsilon$ (while $\eta$ remains fixed). Hence the $O(\epsilon^3)$ term here is essentially fixed too.

We have just demonstrated that this problem has no $C^2$ minimizer ($x^* = t$ is the only candidate, and it would be a $\mathcal{D}^1$ minimizer too if it did minimize over $C^2$), and we can actually show that there is no $\mathcal{D}^1$ minimizer either. To do this, we construct the family of admissible $\mathcal{D}^1$ functions

$$x_\epsilon(t) = \begin{cases} -\epsilon t & \text{if } t \in [0, 0.5], \\ 2(1 + 0.5\epsilon)(t - 1) + 1 & \text{if } t \in [0.5, 1], \end{cases}$$

graphed for $\epsilon = 0.1$ in Figure B.6, whose integral value is

$$J[x_\epsilon] = -\frac{1}{\epsilon(2 + \epsilon)}.$$

In the limit, we get

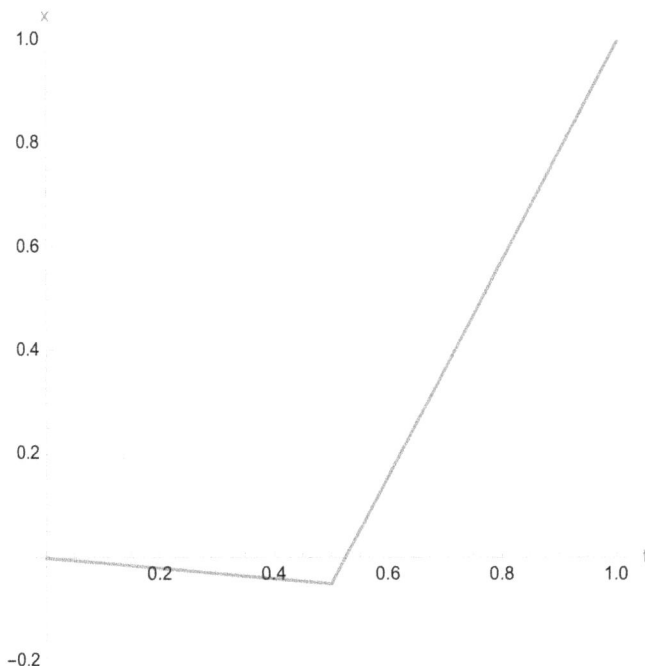

**Figure B.6:** Graph of $x_{0.1}(t)$.

$$\lim_{\epsilon \to 0} J[x_\epsilon] = \lim_{\epsilon \to 0}\left(-\frac{1}{\epsilon(2+\epsilon)}\right) = -\infty,$$

which indicates that the integral value can be made arbitrarily small with functions in this family.

**Problem 6.1 (solution)**

We have already seen where the first endpoint condition leads us. If we instead enforce the other endpoint condition $x(1) = 1$, then we get the field of extremals $x(t) = (1-b)t+b$ for the region $R$ equal to the $tx$-plane minus the vertical line $t = 1$, as shown in Figure B.7. Notice that enforcing an endpoint condition always generates a family of functions that all cover the corresponding point in the $tx$-plane, because all members of the family will necessarily satisfy the endpoint condition. A more fruitful route is to match the slope of the extremal $x^*(t) = t$ to get the field of extremals $\mathcal{F} = \{t + b : b \in \mathbb{R}\}$ for the entire $tx$-plane shown in Figure B.8. Note that simply adding $b$ to our star extremal is a way to generate a field of extremals $\mathcal{F} = \{x^*(t) + b : b \in \mathbb{R}\}$ as long as the functions $x^*(t) + b$ all still satisfy the Euler–Lagrange equation. This will be the case whenever the Euler–Lagrange equation contains only derivatives $\dot{x}$ and/or $\ddot{x}$ and not $x$ itself. An example where this does not work is when the integrand is $f(t, x, \dot{x}) = x^2 + \dot{x}^2$, in which case the Euler–Lagrange equation simplifies to $\ddot{x} = x$.

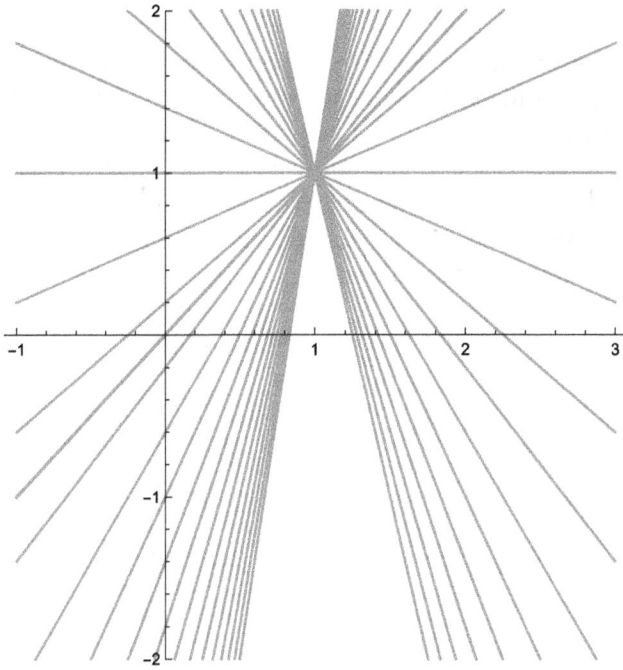

**Figure B.7:** Field of extremals.

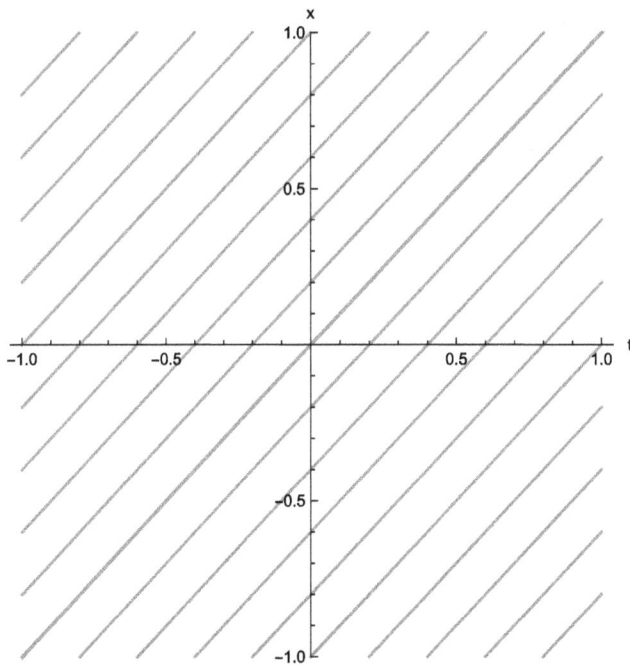

**Figure B.8:** Field of extremals $\mathcal{F} = \{t + b : b \in \mathbb{R}\}$.

**Problem 6.2 (solution)**

The Euler–Lagrange equation in this case simplifies to $\ddot{x} = 0$ with straight-line extremals $x(t) = at + b$. The extremal satisfying the endpoint conditions in this case is $x^*(t) = 2t - 1$. Two obvious fields of extremals are $\mathcal{F}_1 = \{at - 1 : a \in \mathbb{R}\}$ and $\mathcal{F}_2 = \{2t + b : b \in \mathbb{R}\}$, and they both simply cover the region of integration (the first does not simply cover the $x$-axis, but that is excluded from the region of integration). It is easy to compute $f_{\dot{x}\dot{x}} = 2$, so that $x^*(t) = 2t - 1$ passes the strong second-derivative test and is a minimizer.

The excess function is nonnegative in this case since

$$\begin{aligned}
\exc(t, x, \dot{x}, p) &= f(t, x, \dot{x}) - f(t, x, p) - (\dot{x} - p)f_{\dot{x}}(t, x, p) \\
&= \dot{x}^2 - p^2 - (\dot{x} - p)2p \\
&= \dot{x}^2 - 2\dot{x}p + p^2 \\
&= (\dot{x} - p)^2 \\
&\geq 0.
\end{aligned}$$

**Problem 7.1 (solution)**

$$\text{Minimize } \int_0^T -x \, dt \quad \text{over } x \in C^2 \text{ on } [0, T]$$

$$\text{with } x(0) = 0, \, x(T) = 0,$$

$$\text{and } \int_0^T \sqrt{1 + \dot{x}^2} \, dt = 100.$$

**Problem 8.1 (solution)**

$$x_0(t + 1) = a_{0,0}x_0(t) + a_{0,1}x_1(t),$$
$$x_1(t + 1) = a_{1,0}x_0(t) + a_{1,1}x_1(t).$$
$$x_2(t + 1) = a_{2,1}x_1(t) + a_{2,2}x_2(t).$$

**Problem 8.2 (solution)**

We know from (8.5) that

$$J_3(\mathbf{u}) = \frac{\lambda^3}{\text{sur}_3(\mathbf{u})} \frac{\text{sur}_3(\mathbf{u})\,\text{fec}_3(\mathbf{u})}{\lambda^4} = \frac{\text{fec}_3(\mathbf{u})}{\lambda} = 0,$$

since $\text{fec}_3(\mathbf{u}) = 0$. Thus $J_3$ is maximized (with maximum value 0) by any $u_3$, so elephants in the last group can choose any value of reproductive effort $0 \leq u_3 \leq 1$ they like without affecting the reproductive value of their species.

It follows from (8.6) that

$$J_2(\mathbf{u}) = \frac{1}{\lambda}\left(\mathrm{fec}_2(\mathbf{u}) + \frac{\mathrm{sur}_3(\mathbf{u})}{\mathrm{sur}_2(\mathbf{u})}J_3(\mathbf{u})\right) = \frac{\mathrm{fec}_2(\mathbf{u})}{\lambda} = \frac{7u_2 - 2u_1}{\lambda},$$

which is maximized over $0 \le u_2 \le 1$ at $u_2 = 1$. This results in the maximum value of $J_2(u_0, u_1, 1, u_3) = \frac{7-2u_1}{\lambda}$.

Now it follows from (8.6) that

$$J_1(\mathbf{u}) = \frac{1}{\lambda}\left(\mathrm{fec}_1(\mathbf{u}) + \frac{\mathrm{sur}_2(\mathbf{u})}{\mathrm{sur}_1(\mathbf{u})}J_2(\mathbf{u})\right)$$

$$\Downarrow$$

$$J_1(u_0, u_1, 1, u_3) = \frac{1}{\lambda}\left(7u_1 + \frac{(0.8 - 0.2u_1)(7 - 2u_1)}{0.9\lambda}\right),$$

which is maximized over $0 \le u_1 \le 1$ at $u_1 = 1$ (assuming that $\lambda \ge 1$) with maximum value $\frac{10}{3\lambda^2} + 7\lambda$.

Finally, it follows from (8.6) that

$$J_0(\mathbf{u}) = \frac{1}{\lambda}\left(\mathrm{fec}_0(\mathbf{u}) + \frac{\mathrm{sur}_1(\mathbf{u})}{\mathrm{sur}_0(\mathbf{u})}J_1(\mathbf{u})\right)$$

$$\Downarrow$$

$$J_0(u_0, 1, 1, u_3) = \frac{1}{\lambda}\left(0 + 0.9\left(\frac{10}{3\lambda^2} + 7\lambda\right)\right),$$

which is maximized at any $u_0$. We conclude that the reproductive value of this grouping is maximized at $(u_0, 1, 1, u_3)$ for any $u_0$ and $u_3$ satisfying $0 \le u_0 \le 1$ and $0 \le u_3 \le 1$, with maximum value $\frac{3}{\lambda^3} + 6.3$.

## Problem 9.1 (solution)
$U = [0, M], f_0(x, u) = 1, f(x, u) = ax - u.$

## Problem 9.2 (solution)
The Euler–Lagrange equation in this case gives $\dot{\lambda} = H_x$.

The target curve in this case is $\mathrm{tar}(t) = x_1$, so $\mathrm{tar}'(T^*) = 0$. Thus the transversality condition applied to $\lambda \cdot \dot{x} + H(\lambda, x, u)$ gives

$$(\lambda(T^*) \cdot \dot{x}^*(T^*) + H(T^*)) - \dot{x}^*(T^*) \cdot \lambda(T^*) = 0 \quad \Longleftrightarrow \quad H(T^*) = 0.$$

## Problem 9.3 (solution)
1.

$$H(\lambda, x, u) = 1 - \lambda \cdot (ax - u).$$

2. $\dot{\lambda}(t) = -a\lambda$ implies $\lambda^*(t) = c_1 e^{-at}$ for some constant $c_1$. It follows that

$$H(\lambda^*, x, u) = 1 - c_1 e^{-at}(ax - u),$$

so Pontryagin's principle gives

$$u^*(t) = \begin{cases} 0 & \text{if } c_1 \geq 0, \\ M & \text{otherwise,} \end{cases} \tag{B.5}$$

which means that the Hamiltonian satisfies

$$H(\lambda^*, x, u) = \begin{cases} 1 - c_1 e^{-at} ax & \text{if } c_1 \geq 0, \\ 1 - c_1 e^{-at}(ax - M) & \text{otherwise.} \end{cases} \tag{B.6}$$

From (B.5) and the state equation $\dot{x} = ax - u$ we have

$$x^*(t) = \begin{cases} c_2 e^{at} & \text{if } c_1 \geq 0, \\ c_3 e^{at} + \frac{M}{a} & \text{otherwise,} \end{cases}$$

which we will now refine since it contains various arbitrary constants.

3. The first endpoint condition $x(0) = x_0$ gives

$$x^*(t) = \begin{cases} x_0 e^{at} & \text{if } c_1 \geq 0, \\ (x_0 - \frac{M}{a})e^{at} + \frac{M}{a} & \text{otherwise.} \end{cases}$$

The second endpoint condition $x(T^*) = x_1$ then leads to the equations

$$x_1 = x^*(T^*) = \begin{cases} x_0 e^{aT^*} & \text{if } c_1 \geq 0, \\ (x_0 - \frac{M}{a})e^{aT^*} + \frac{M}{a} & \text{otherwise.} \end{cases}$$

We now solve these for $T^*$ and see what conclusions we can draw.

 — Case $c_1 \geq 0$ (with $u^* = 0$): we get

$$T^* = \frac{\ln(\frac{x_1}{x_0})}{a},$$

which must be nonnegative to make sense as an end time. Since $a > 0$ and since the natural log is nonnegative only for inputs at or above 1, we conclude that $\frac{x_1}{x_0} \geq 1$ which simplifies to $x_1 \geq x_0$.

 — Case $c_1 < 0$ (with $u^* = M$): we get

$$T^* = \frac{\ln(\frac{x_1 - \frac{M}{a}}{x_0 - \frac{M}{a}})}{a},$$

which must be nonnegative to make sense as an end time. Since $\alpha > 0$ and since the natural log is nonnegative only for inputs at or above 1, we conclude that

$$1 \le \frac{x_1 - \frac{M}{\alpha}}{x_0 - \frac{M}{\alpha}}. \tag{B.7}$$

If the denominator $x_0 - \frac{M}{\alpha}$ is positive, then we can multiply it on both sides of (B.7):

$$x_0 - \frac{M}{\alpha} > 0 \quad \& \quad x_0 - \frac{M}{\alpha} \le x_1 - \frac{M}{\alpha}$$

$$\Downarrow \tag{B.8}$$

$$\frac{M}{\alpha} < x_0 \le x_1.$$

If instead the denominator $x_0 - \frac{M}{\alpha}$ is negative, then multiplying it on both sides of (B.7) switches the inequality:

$$x_0 - \frac{M}{\alpha} < 0 \quad \& \quad x_0 - \frac{M}{\alpha} \ge x_1 - \frac{M}{\alpha}$$

$$\Downarrow \tag{B.9}$$

$$\frac{M}{\alpha} > x_0 \ge x_1.$$

Transversality $H(T^*) = 0$ applies in this case, which leads via (B.6) to

$$0 = H(T^*) = \begin{cases} 1 - c_1 e^{-\alpha T^*} \alpha x_1 & \text{if } c_1 > 0, \\ 1 - c_1 e^{-\alpha T^*} (\alpha x_1 - M) & \text{if } c_1 < 0, \end{cases}$$

where we have applied the endpoint condition $x^*(T^*) = x_1$ and we have ruled out $c_1 = 0$ from the $c_1 \ge 0$ case since that would evidently violate $H(T^*) = 0$.

- Case $c_1 > 0$ (with $u^* = 0$): we get $1 = c_1 e^{-\alpha T^*} \alpha x_1$, from which we can conclude that $x_1 > 0$ (since $c_1 e^{-\alpha T^*} \alpha > 0$ and its product with $x_1$ needs to match the positive number 1).

- Case $c_1 < 0$ (with $u^* = M$): we get $1 = c_1 e^{-\alpha T^*} (\alpha x_1 - M)$, from which we can conclude that $\alpha x_1 < M$. Note that if this is not the case, then we can never control the infection to the target level (since the natural growth overwhelms our maximum ability to vaccinate). Since the bound $\alpha x_1 < M$ conflicts with (B.8), we conclude that (B.9) must hold for optimal control $u^* = M$ to work.

The final optimal control is thus

$$u^*(t) = \begin{cases} 0 & \text{if } x_1 \ge x_0, \\ M & \text{if } \frac{M}{\alpha} > x_0 \ge x_1, \\ \text{none} & \text{otherwise,} \end{cases}$$

with corresponding optimal end time

$$
T^* = \begin{cases}
\dfrac{\ln\left(\frac{x_1}{x_0}\right)}{a} & \text{if } x_1 \geq x_0, \\[3mm]
\dfrac{\ln\left(\frac{x_1 - \frac{M}{a}}{x_0 - \frac{M}{a}}\right)}{a} & \text{if } \frac{M}{a} > x_0 \geq x_1, \\[3mm]
\text{none} & \text{otherwise.}
\end{cases}
$$

The interpretation is that no vaccination is done if the target infection level is above the current infection level (a dystopian scenario) and that the maximum vaccination rate is used in the alternate scenario, with the caveat that even the maximum vaccination rate will not be enough to drive the number of infections to $x_1$ if the initial number of infections $x_0$ is greater than or equal to $\frac{M}{a}$.

This optimal control $u^*$ and end time $T^*$ constitute the solution to the problem and happen to result in an optimal state of

$$
x^*(t) = \begin{cases}
x_0 e^{at} & \text{if } x_1 \geq x_0, \\[2mm]
\left(x_0 - \frac{M}{a}\right)e^{at} + \frac{M}{a} & \text{if } \frac{M}{a} > x_0 \geq x_1, \\[2mm]
\text{none} & \text{otherwise.}
\end{cases}
$$

### Problem 11.1 (solution)

1. The Hamiltonian in this case is

$$
H(\lambda, \mathbf{x}, u) = 1 - \lambda_1 x_2 - \lambda_2 u,
$$

where we have used two costates $\lambda_1$ and $\lambda_2$ (one for each component state equation).

2. – the costate equations $\dot{\lambda}_i = \frac{\partial H}{\partial x_i}$ in this case become

$$
\begin{aligned}
\dot{\lambda}_1 &= 0 & \implies & \quad \lambda_1(t) = c_1, \\
\dot{\lambda}_2 &= -\lambda_1 & \implies & \quad \lambda_2(t) = c_2 - c_1 t
\end{aligned}
$$

in terms of arbitrary constants $c_1$ and $c_2$.

– Pontryagin's principle minimizes the Hamiltonian

$$
1 - \lambda_1 x_2 - \lambda_2 u = 1 - c_1 x_2 - (c_2 - c_1 t)u
$$

with respect to $u$ (with $|u(t)| \leq 1$), which leads to the optimal control $u^*(t) = \text{sgn}(s(t))$ in terms of the switching function $s(t) = c_2 - c_1 t$. There is at most one switch since this (linear) switching function has at most one zero.

– the state equations $\dot{x}_1 = x_2$ and $\dot{x}_2 = u^*$ then lead to

$$
x_1^*(t) = \text{sgn}(s(t))\left(\frac{t^2}{2}\right) + c_3 t + c_4,
$$

$$x_2^*(t) = \text{sgn}(s(t))t + c_3.$$

3. The initial condition $\left[\begin{smallmatrix} x_1(0) \\ x_2(0) \end{smallmatrix}\right] = \left[\begin{smallmatrix} a \\ b \end{smallmatrix}\right]$ gives us

$$x_1^*(t) = \text{sgn}(s(t))\left(\frac{t^2}{2}\right) + bt + a,$$
$$x_2^*(t) = \text{sgn}(s(t))t + b,$$

## Problem 12.1 (solution)

Multiplying both sides of $0 = s(t) = c_3 e^{-\epsilon_1 t} + c_4 e^{-\epsilon_2 t}$ by $e^{\epsilon_1 t}$ and solving for $t$ give

$$t = \frac{\ln(\frac{-c_3}{c_4})}{\epsilon_1 - \epsilon_2},$$

so there is at most one switch.

## Problem 15.1 (solution)

(a) Similarly to $m_1$, we get $p_1 = e^{-0.2\pi} + 1$.

(b) We get to $(m_2, 0)$ by backtracking $\pi$ time units from $(p_1, 0)$ (using $u^* = -1$). Therefore we can use the $u^* = -1$ state trajectory forward in $\pi$ time units from $(m_2, 0)$ to $(p_1, 0)$ to get

$$p_1 = x_1^*(\pi) = c_1 e^{0.2\pi} \cos(\pi) - 1 = -c_1 e^{0.2\pi} - 1.$$

Substituting our preceding formula $p_1 = e^{-0.2\pi} + 1$ and solving for $c_1$, we get

$$c_1 = -e^{(-0.2\pi)(2)} - 2e^{-0.2\pi}.$$

The exact value for $m_2$ follows:

$$m_2 = x_1^*(0) = c_1 - 1 = -e^{(-0.2\pi)(2)} - 2e^{-0.2\pi} - 1.$$

## Problem 16.1 (solution)

Costate perpendicularity in this case implies that $\lambda_1(T^*) = 0$ if the target curve is hit when $x_1 > 0$, in which case the costates are $\lambda_1(t) = 0$ and $\lambda_2(t) = c_4$, and the switching function is $s(t) = \lambda_2(t) = c_4$. Thus there are no switches for optimal state trajectories hitting the target at $x_1 > 0$.

If the target curve is instead hit at $x_1 = 0$, then the costate perpendicularity condition says nothing since all vectors are "perpendicular" to the endpoint of the target curve at $x_1 = 0$ (since at this endpoint the target curve has tangent lines of all slopes). In this case the switching function is $s(t) = \lambda_2(t) = c_4 - c_3 t$, which means at most one switch. The

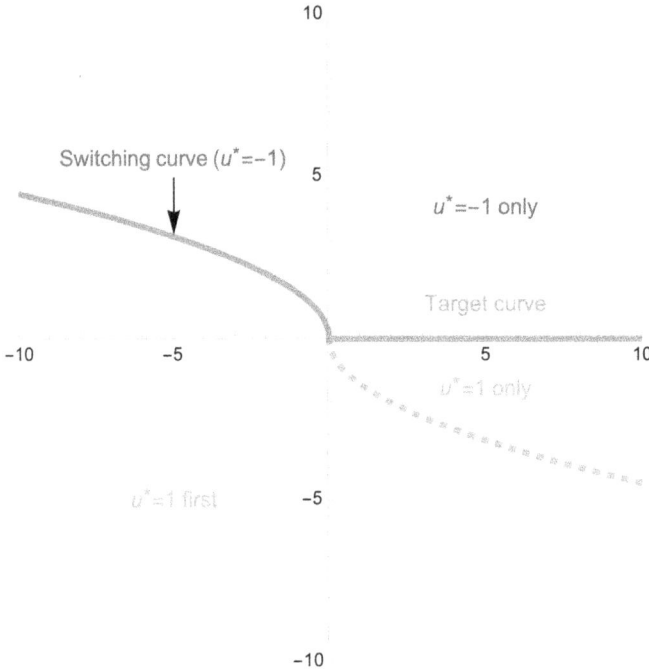

**Figure B.9:** Optimal synthesis for Problem 16.1.

target curve, switching curve, and optimal control synthesis are pictured in Figure B.9. Any initial point below the $u^* = -1$ part of the switching curve joined to the dashed curve has optimal control of $u^* = 1$ until switching onto the switching curve with $u^* = -1$. Any initial point on or above the dashed curve and below the $x_1$-axis has optimal control $u^* = 1$. Any initial point above the switching curve and above the $x_1$-axis has optimal control $u^* = -1$.

## Problem 17.1 (solution)
The cost functional is $\int_0^T 1\,dt$, and the state equations are

$$\dot{x}_1 = u_1 \quad \text{and} \quad \dot{x}_2 = u_2,$$

so the Hamiltonian is

$$H(\lambda, \mathbf{x}, \mathbf{u}) = 1 - \lambda_1 u_1 - \lambda_2 u_2.$$

Thus we have costate equations

$$\dot{\lambda}_1 = 0 \quad \Longrightarrow \quad \lambda_1 = c_1,$$
$$\dot{\lambda}_2 = 0 \quad \Longrightarrow \quad \lambda_2 = c_2,$$

and optimal controls

$$u_1^* = \text{sgn}(\lambda_1) = \text{sgn}(c_1),$$
$$u_2^* = \text{sgn}(\lambda_2) = \text{sgn}(c_2).$$

In particular, this means that we can treat the optimal controls as constants when we solve the state equations with $x_1(0) = a$ and $x_2(0) = b$ to get

$$x_1^*(t) = u_1^* t + a \quad \text{and} \quad x_2^*(t) = u_2^* t + b.$$

Applying the endpoint condition, we get

$$0 = x_1^*(T^*) = u_1^* T^* + a \quad \Longrightarrow \quad T^* = \frac{-a}{u_1^*},$$

$$0 = x_2^*(T^*) = u_2^* T^* + b \quad \Longrightarrow \quad T^* = \frac{-b}{u_2^*}.$$

From this and the fact that $T^* \geq 0$ we conclude that

$$u_1^* = -\text{sgn}(a) \quad \text{and} \quad u_2^* = -\text{sgn}(b)$$

and that $|a| = |b|$ to have a consistent end time $T^* = |a| = |b|$.

Notice that this solution only works for initial states $(a, b)$ satisfying $|a| = |b|$; however, our intuition tells us that we should be able to optimally control all initial states, not just those on the two crossing lines in Figure B.10. The trouble is that we neglected the important situation here where one of the switching functions $s_1(t) = c_1$ or $s_2(t) = c_2$ equals zero.

### Problem 17.2 (solution)
The first optimal control $u_1^* = -\text{sgn}(a)$ is already the same as we found in the nonsingular case, so we focus on the second optimal control $u_2^*$. Since $|a| = |b|$, the defining integral (17.2) amounts to

$$\int_0^{|b|} u_2^*(\tau) \, d\tau = -b,$$

which can only be achieved with $|u_2^*(t)| \leq 1$ by the choice $u_2^* = -1$ (when $b > 0$) or $u_2^* = 1$ (when $b < 0$). This gives the same second optimal control $u_2^* = -\text{sgn}(b)$ as we found in the nonsingular case.

### Problem 17.3 (solution)
This initial state satisfies the conditions for singular case 2, so the optimal controls are $u_1^* = -1$ and any $u_2^*$ satisfying $|u_2^*(t)| \leq 1$ and

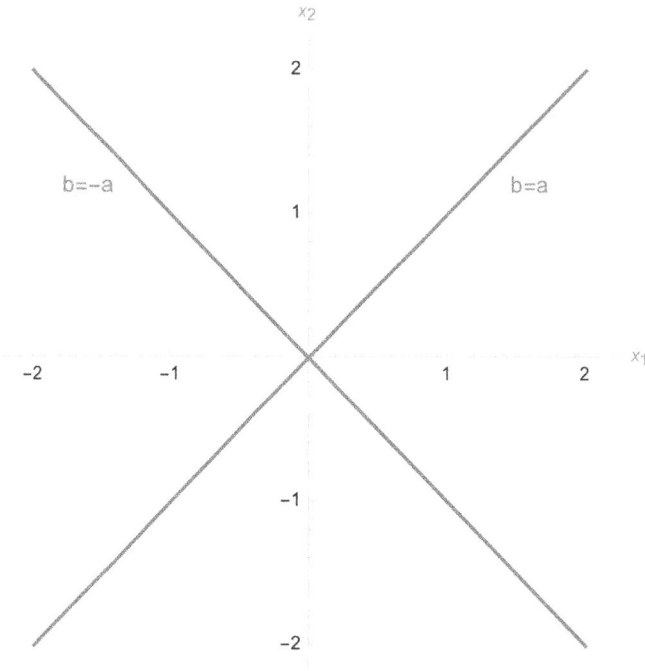

**Figure B.10:** Initial states $(a, b)$ satisfying $|a| = |b|$.

$$\int_0^2 u_2^*(\tau)\, d\tau = -1.$$

Some possibilities for $u_2^*$ are

$$u_2^*(t) = -\frac{1}{2} \quad \text{or} \quad u_2^*(t) = -\frac{t}{2} \quad \text{or} \quad u_2^*(t) = -\frac{\pi}{4} \sin\left(\frac{\pi}{2} t\right).$$

Figure B.11 shows the optimal trajectories associated with these and two other singular optimal control choices.

**Problem 18.1 (solution)**
Pontryagin's principle in this case works out a little differently since minimizing $H$ with respect to $u$ amounts to minimizing $|u| - \lambda_2 u$:

$$u^*(t) = \begin{cases} 0 & \text{if } |\lambda_2(t)| < 1, \\ \text{sgn}(\lambda_2(t)) & \text{if } |\lambda_2(t)| > 1, \\ \in [0, 1] & \text{if } \lambda_2(t) = 1 \text{ (singular)}, \\ \in [-1, 0] & \text{if } \lambda_2(t) = -1 \text{ (singular)}. \end{cases}$$

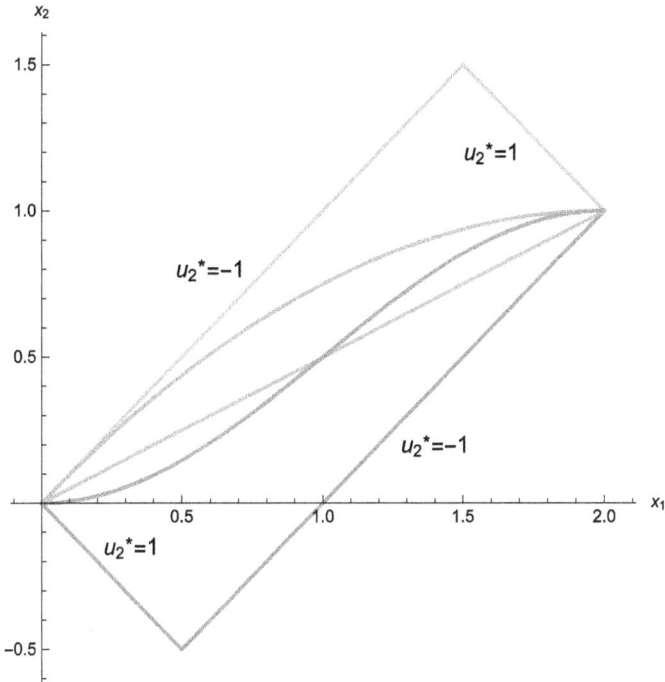

**Figure B.11:** Possible optimal trajectories.

Note that the singular cases $\lambda_2(t) = c_2 - c_1 t = \pm 1$ happen when $c_1 = 0$ and $c_2 = \pm 1$, so we cannot rule them out since they have

$$H(\lambda, \mathbf{x}, u) = -\lambda_1 x_2 = -c_1 x_2 = 0,$$

which clearly satisfies $H(T^*) = 0$.

## Problem 19.1 (solutions)

The Hamiltonian in this case is

$$H(\lambda, \mathbf{x}, u) = \tau + |u| - \lambda_1 x_2 - \lambda_2 u,$$

so the costate equations are

$$\dot{\lambda}_1 = 0 \quad \Longrightarrow \quad \lambda_1 = c_1,$$
$$\dot{\lambda}_2 = -\lambda_1 = -c_1 \quad \Longrightarrow \quad \lambda_2 = c_2 - c_1 t.$$

Just as with energy-optimal control, Pontryagin's principle gives

$$u^* = \begin{cases} 0 & \text{if } |\lambda_2(t)| < 1, \\ \operatorname{sgn}(\lambda_2) & \text{if } |\lambda_2(t)| > 1, \\ \in [0,1] & \text{if } \lambda_2(t) = 1 \text{ (singular)}, \\ \in [-1,0] & \text{if } \lambda_2(t) = -1 \text{ (singular)}. \end{cases}$$

Thus the singular cases can only happen when $c_1 = 0$ and $c_2 = \pm 1$. However, this implies that

$$H(\lambda, \mathbf{x}, u) = \tau - \lambda_1 x_2 = \tau - c_1 x_2 = \tau,$$

which is always positive and so violates $H(T^*) = 0$. Recall that this was not the case for energy-optimality (essentially, when $\tau = 0$).

## Problem 19.2 (solutions)

a) As with the case we already covered, we get the constant vertical component $b_0 = \frac{\tau}{c_1}$ in the $x_1 x_2$-plane of the entire (horizontal) trajectory associated with $u^* = 0$. Now though we are switching from $u^* = -1$ to $u^* = 0$ and the switch occurs at negative values of $b_0$. Thus the time $\frac{2}{|c_1|}$ spent on the $u^* = 0$ trajectory translates via $c_1 = \frac{\tau}{b_0}$ into $\frac{2|b_0|}{\tau} = \frac{-2b_0}{\tau}$. Following the same argument, we get the terminal point

$$\left( a_0 + b_0 \frac{-2b_0}{\tau}, b_0 \right)$$

for the $u^* = 0$ trajectory. Using this point as the initial state and solving the state equations $\dot{x}_1 = x_2$ and $\dot{x}_2 = 1$ with $u^* = 1$, we get the parameterization

$$x_1^*(t) = a_0 + b_0 \frac{-2b_0}{\tau} + b_0 t + \frac{t^2}{2} \quad \text{and} \quad x_2^*(t) = b_0 + t$$

of the $u^* = 1$ switching curve to the target state $(0, 0)$. Evidently, this hits $(0, 0)$ at $t = -b_0$ (when $x_2^*(t) = 0$), so we conclude that

$$0 = x_1^*(-b_0) = a_0 + b_0 \frac{-2b_0}{\tau} - b_0 b_0 + \frac{b_0^2}{2}$$

$$\Downarrow$$

$$a_0 = b_0 \frac{2b_0}{\tau} + \frac{b_0^2}{2} = \left( \frac{1}{2} + \frac{2}{\tau} \right) b_0^2,$$

and we have our result.

b) The formulas

$$x_1 = -\left( \frac{1}{2} + \frac{2}{\tau} \right) x_2^2 \quad \text{and} \quad x_1 = \left( \frac{1}{2} + \frac{2}{\tau} \right) x_2^2$$

defining the $u^* = 0$ part of the switching curve approach

$$x_1 = -\left(\frac{1}{2}\right)x_2^2 \quad \text{and} \quad x_1 = \left(\frac{1}{2}\right)x_2^2$$

as $\tau \to \infty$. These are precisely the $u^* = -1$ and $u^* = 1$ time-optimal switching curves in nonparametric form.

c)   The formulas

$$x_1 = -\left(\frac{1}{2} + \frac{2}{\tau}\right)x_2^2 \quad \text{and} \quad x_1 = \left(\frac{1}{2} + \frac{2}{\tau}\right)x_2^2$$

defining the $u^* = 0$ part of the switching curve approach $x_1 = -\infty$ and $x_1 = \infty$ as $\tau \to 0$, which means the $u^* = 0$ part of the switching curve approaches the $x_1$-axis. This is exactly what we saw in energy-optimality where there was no optimal control using $u^* = 1 \to 0 \to -1$ or $u^* = -1 \to 0 \to 1$, since the optimal switch to $u^* = 0$ wants to happen on the $x_1$-axis (where no progress is made toward $(0,0)$).

## Solutions to lab activities

### Lab Activity 1 (solutions)
1.

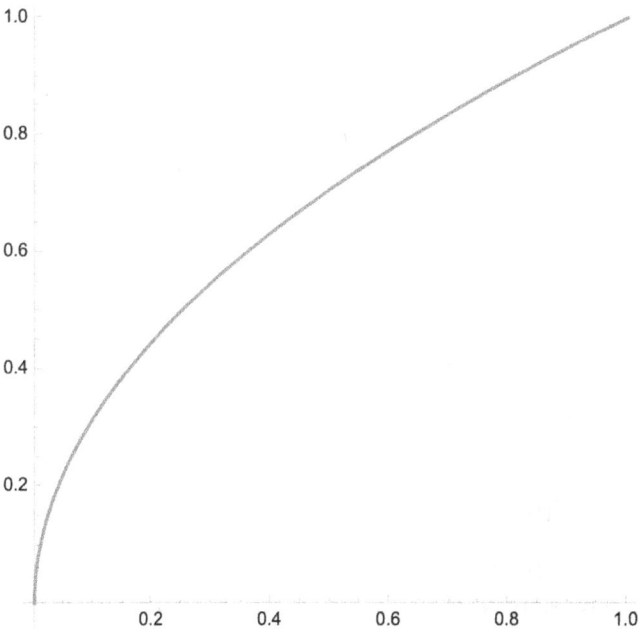

2.

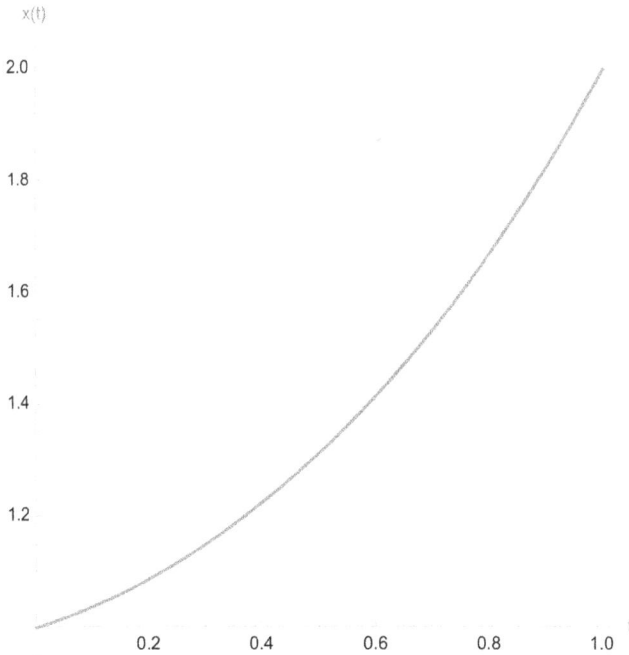

**Lab Activity 2 (solutions)**

1.  The Euler–Lagrange equation in this case is

$$x(t)(\ddot{x}(t)x(t) + \dot{x}(t) - 1) = 0,$$

which is equivalent to

$$x(t) = 0 \quad \text{or} \quad \ddot{x}(t) = \frac{1 - \dot{x}(t)^2}{x(t)}.$$

We can observe the solutions $x^*(t) = 0$ (from the first equation) or $x^*(t) = t$ (from the second equation, since linear with slope 1 gives zero on both sides of that differential equation). To determine other solutions to the second (differential) equation, we can use Mathematica (but not Python in this case) to obtain

$$x^*(t) = \pm\sqrt{t^2 + 2c_2 t - e^{2c_1} + c_2^2}.$$

(a)  $d = 2$: the extremal is $x^*(t) = t$ on $[0, 2]$.

(b)  $d = 1$: the presumptive extremal is $x^*(t) = \frac{\sqrt{t(2t-3)}}{\sqrt{2}}$, which is not continuous on $[0, 2]$, so no extremal exists in this case.

(c)  $d = 0$: the extremal is $x^*(t) = 0$.

2.  The integrand is nonnegative, so the integral functional always produces nonnegative values. The extremals $x^*(t) = t$ (in case (a)) and $x^*(t) = 0$ (in case (c)) both have the integral value equal to zero, so they are solutions to their respective problems. These are both $C^2$ functions, so they also are automatically $\mathcal{D}^1$ functions. For case (b), there is a $\mathcal{D}^1$ minimizer using $x = 0$, then a straight-line piece of slope 1:

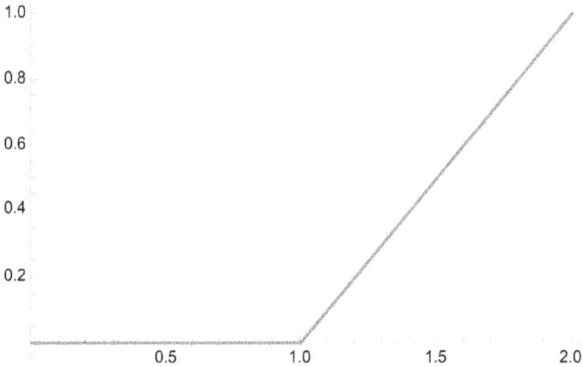

### Lab Activity 3 (solutions)

Assuming that $x^*(t) \geq 0$, we get the extremals $x^*(t) = \sqrt{2t\lambda - t^2}$. We then solve the (hits-the-target) endpoint condition $x^*(T^*) = 0$ to obtain the relationship $T^* = 2\lambda$ (ruling out $T^* = 0$, which would not use any rope). We would like to solve the integral constraint for $\lambda$ (from 0 to $2\lambda$), but that cannot be done explicitly. Instead, we graph the integral for a range of $\lambda$ to estimate the value of $\lambda$ where the integral equals 100. This gives us $\lambda \approx 31.83$, from which we deduce that $T^* = 2(31.83) = 63.66$. We then graph $x^*(t) = \sqrt{63.66t - t^2}$

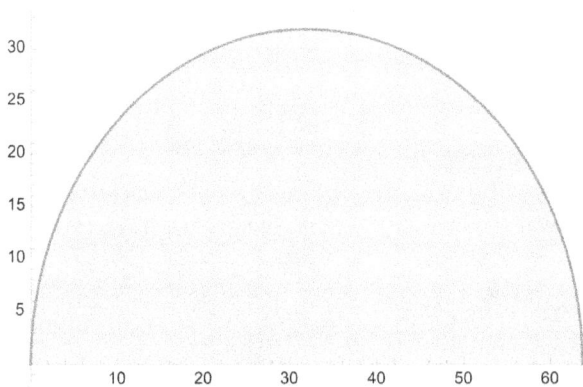

and compute $\int_0^{63.66} x^*(t)\, dt = 1591.45$.

Another way to figure out that Dido's solution is a half-circle is to square the extremal $x^*(t) = \sqrt{2t\lambda - t^2}$ and notice (by completing the square) that it satisfies

$$x^*(t)^2 + (t - \lambda)^2 = \lambda^2,$$

which means that the graph of $x^*$ is on the circle of radius $\lambda$ about the point $(\lambda, 0)$ in the $tx$-plane.

### Lab Activity 4 (solutions)

1.  i)  Any $(a, b)$ on the switching curve.

    ii), iii), and iv):

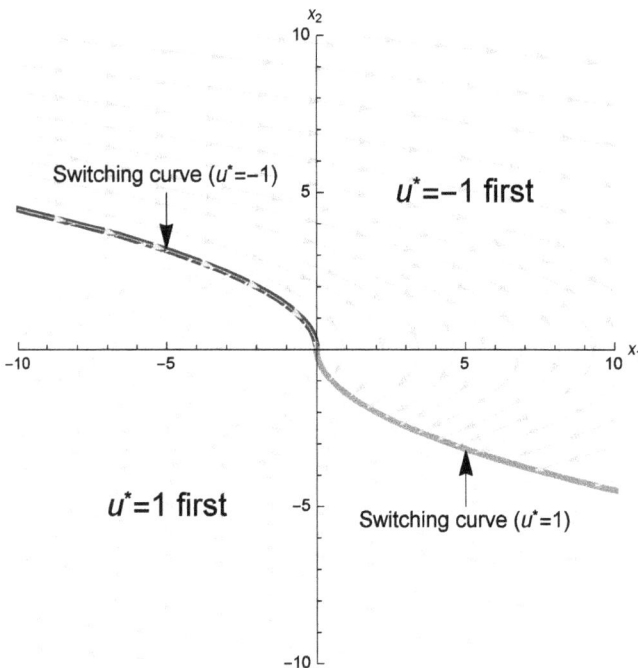

2.  Using the initial point $(0, 0)$ (which is in fact the terminal point of each separate part of the switching curve), we get

$$(sc_1(s), sc_2(s)) = \begin{cases} (-\frac{s^2}{2}, -s) & \text{for } u^* = -1, \\ (\frac{s^2}{2}, s) & \text{for } u^* = 1, \end{cases}$$

parameterized in backward $s$-time by $s \in (-\infty, 0]$. An alternate form substitutes $-s$ for $s$ and uses forward $s$-time $s \in [0, \infty)$:

$$(sc_1(s), sc_2(s)) = \begin{cases} (-\frac{s^2}{2}, s) & \text{for } u^* = -1, \\ (\frac{s^2}{2}, -s) & \text{for } u^* = 1. \end{cases}$$

## Lab Activity 5 (solutions)

1.

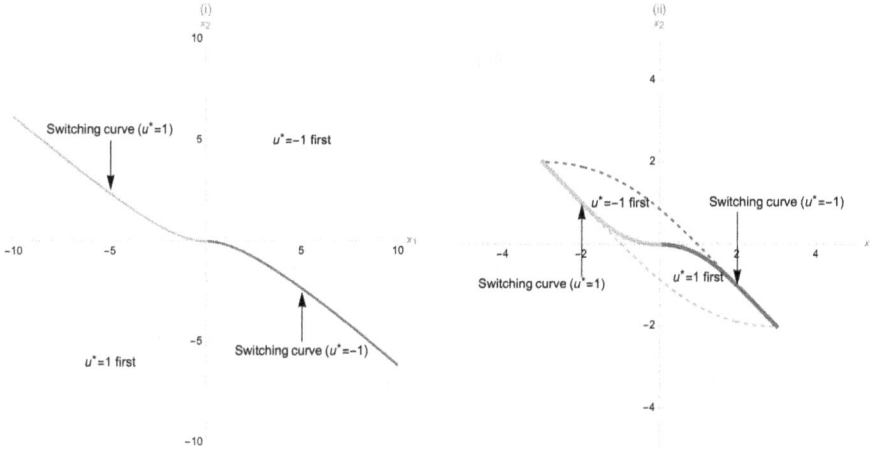

2.   The equilibria in system (i) are $(3, 2)$ (when $u^* = 1$) and $(-3, -2)$ (when $u^* = -1$). The eigenvalues of $A$ are both negative ($-5$ and $-1$), so these equilibria are sinks. The equilibria in system (ii) are $(-3, 2)$ (when $u^* = 1$) and $(3, -2)$ (when $u^* = -1$). The eigenvalues of $A$ are both positive ($5$ and $1$), so these equilibria are sources.

## Lab Activity 6 (solutions)

1.   The equilibria are saddles since the eigenvalues of the matrix are $4$ and $-2$. The switching curve is given by

$$sc(s) = \begin{cases} \overbrace{(-1 + e^s(\cosh(3s) + 2\sinh(3s)), \frac{1}{3}(-4 + e^{-2s} + 3e^{4s}))}^{u^* = -1} \\ \underbrace{(\frac{1}{3}(2 + e^{-2s} - 3e^{4s}), 2 - e^s(2\cosh(3s) + \sinh(3s)))}_{u^* = 1} \end{cases}$$

parameterized by $s \in (-\infty, 0]$, and the region of controllability is

$$\{(x_1, x_2) \text{ such that } -x_1 - 3 < x_2 < -x_1 + 3\}:$$

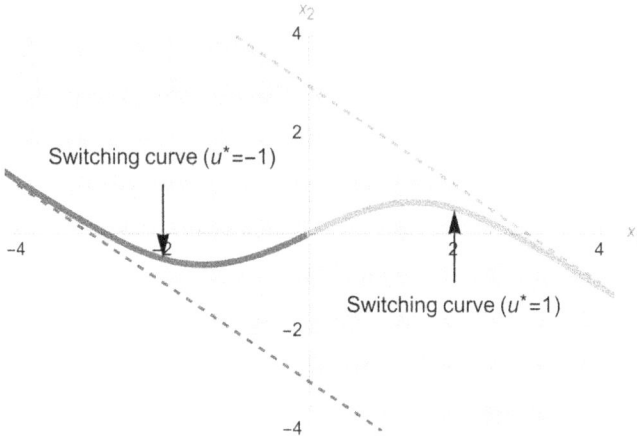

Note that we can get the formula for the region of controllability in at least two different ways:

i)  By taking the limit as $s \to -\infty$ of the ratio $x_2(s)/x_1(s)$ (where $x_1(s)$ and $x_2(s)$ parameterize each part of the switching curve) to find the slope $m$ of the (evidently linear) asymptote $x_2 = mx_1 + b$. This results in $m = -1$ for both cases of optimal control. We then get the intercepts $b$ in each case by solving for $b = x_2(s) - mx_1(s) = x_2(s) + x_1(s)$ and taking the limit as $s \to -\infty$.

ii) By noticing that each boundary line passes through a corresponding equilibrium point and follows the direction of an eigenvector. For example, the upper boundary line passes through the $u^* = 1$ equilibrium at $(1, 2)$ and is in the direction of the eigenvector $(-1, 1)$. Thus the equation for the line has slope $-1$ (the slope of the eigenvector) and passes through $(1, 2)$. Using the point-slope formula, we get

$$x_2 - 2 = -(x_1 - 1) \quad \Longrightarrow \quad x_2 = -x_1 + 3,$$

as above.

2.  The equilibria are degenerate since the eigenvalues of the matrix are both zero. Since the state equations here are the same as those for the original nanosatellite problem, Pontryagin's principle again requires us to minimize the Hamiltonian

$$1 - \lambda_1 x_2 - \lambda_2 u = 1 - c_1 x_2 - (c_2 - c_1 t)u$$

with respect to $u$. However, now we use the bounds $-1 \leq u \leq 0$, which leads to an optimal control of

$$u^*(t) = \begin{cases} 0 & \text{if } s(t) > 0, \\ -1 & \text{if } s(t) < 0, \end{cases}$$

in terms of the switching function $s(t) = c_2 - c_1 t$. As before, there is at most one switch since this (linear) switching function has at most one zero.
The switching curve is given by

$$sc(s) = \left(-\frac{s^2}{2}, -s\right) \quad \text{for } u^* = -1$$

parameterized by $s \in (-\infty, 0]$ (there is no part of the switching curve associated with $u^* = 0$), and the region of controllability is

$$\{(x_1, x_2) \text{ such that } x_1 < 0 \text{ and } 0 < x_2 \leq \sqrt{-2x_1}\} \cup \{0, 0\}:$$

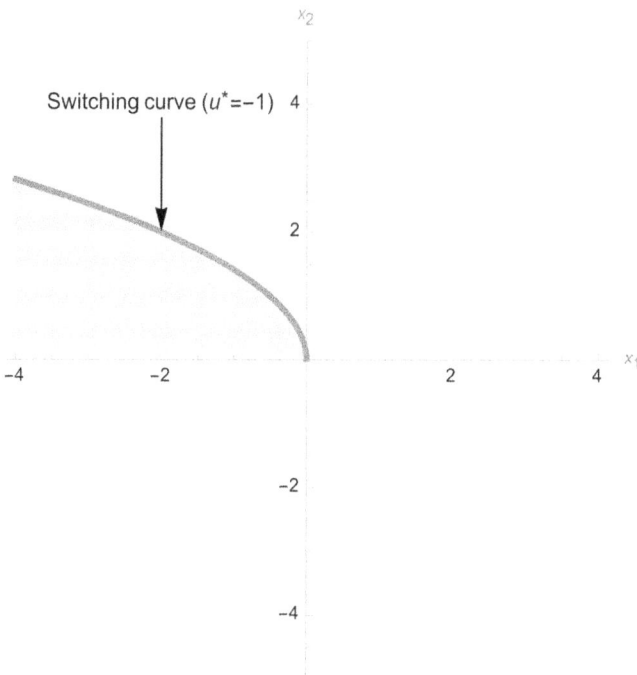

Note that we get the formula for the region of controllability from the parameterization of the switching curve by solving $x_1 = -\frac{s^2}{2}$ for $s = -\sqrt{-2x_1}$ (we use the negative since we are using backward time) and substituting that into $x_2 = -s$.

**Lab Activity 7 (solutions)**
1. $a = 0$: The switching curve in this case looks like

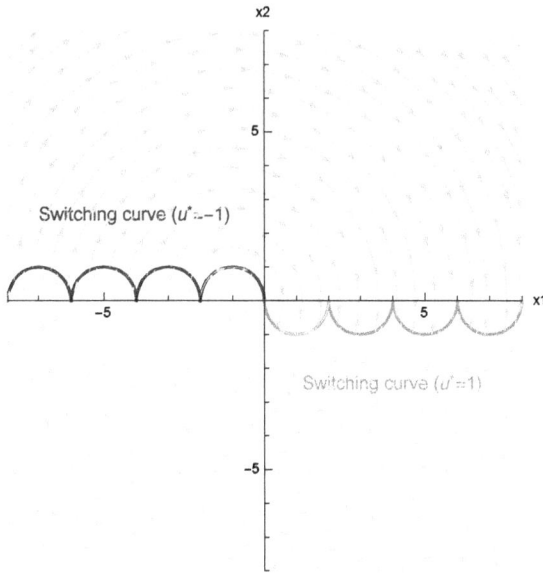

and continues in the same pattern along the $x_1$-axis. Notice that only the semicircles through the origin $(0,0)$ represent trajectories of the system. The other semicircles are simply locations in the $x_1x_2$-plane, where a switch is made.

2.  $\alpha = -0.2$: The switching curve in this case looks like

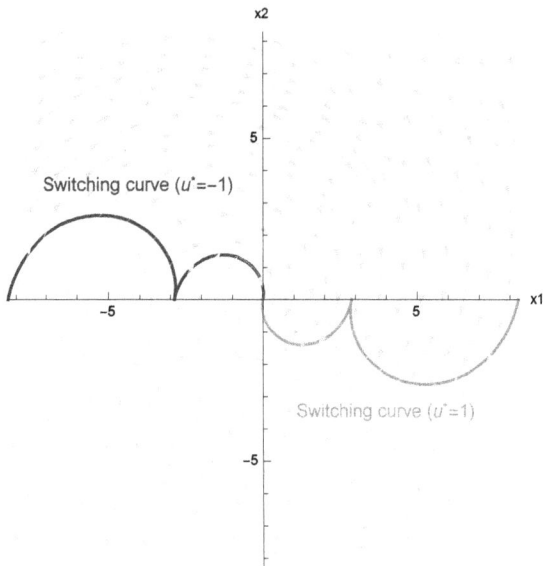

and continues in the same pattern along the $x_1$-axis (expanding in size).

3. $\alpha = 0.2$: The switching curve in this case looks like

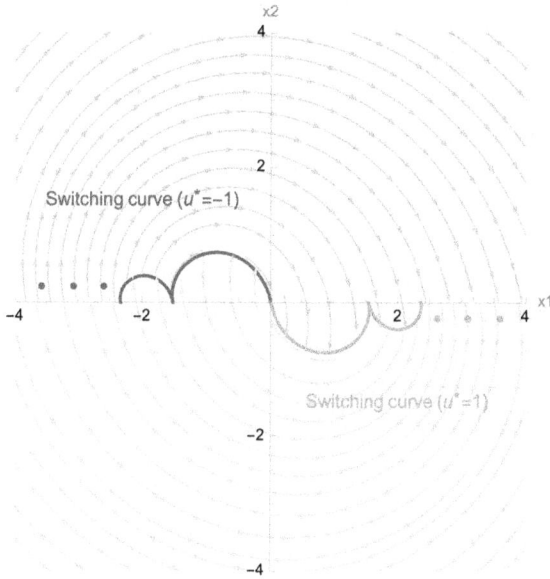

and continues in the same pattern along the $x_1$-axis (shrinking in size).

## Lab Activity 8 (solutions)
1.

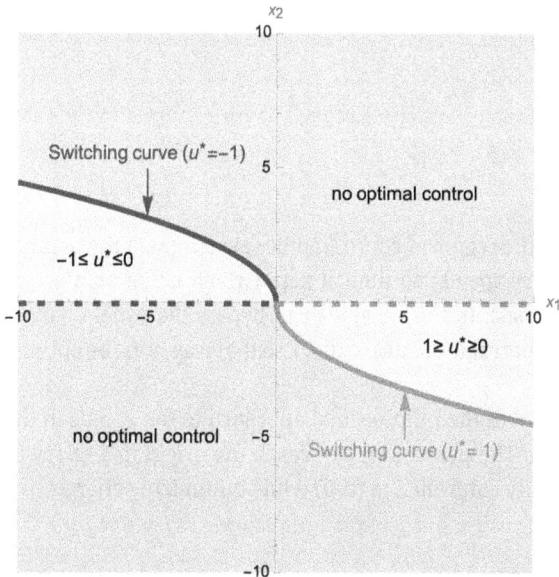

There are infinitely many different optimal state trajectories (including those using singular controls) for initial points between the switching curve and the $x_1$-axis. On the other hand, any initial point outside this region has no optimal control. This follows from the fact that energy-optimality requires minimal distance covered in the $x_2$-direction, so any trajectory crossing the negative (positive) $x_1$-axis would want to switch to a trajectory with $u \in [-1, 0]$ ($u \in [0, 1]$) as soon as possible after crossing. Of course, switching to a trajectory on the $x_1$-axis would not work since those trajectories have zero velocity and so do not move. Thus we can always turn closer and closer to the $x_1$-axis to minimize energy consumption, but we can never turn at the axis itself. Hence there is no optimal control.

2. We get the same optimal control patterns as in the usual energy-optimal control to $(0, 0)$, since the analysis there is essentially unaffected (note that the fixed end time $T^* = 5$ means that we cannot enforce $H(T^*) = 0$). Any initial point $(a, b)$ that can be time-optimally controlled to the origin within 5 seconds can be optimally controlled to be there at 5 seconds while minimizing energy, since the control $u^* = 0$ can always be used at $(0, 0)$ to "stall" there.

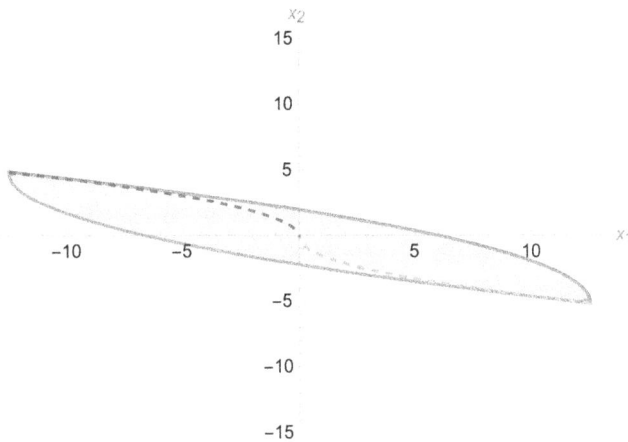

This includes the boundary of the region of controllability even though the optimal control for states initiated there spends no time at zero (which is not an allowed pattern according to our analysis). In this case, we can bypass the usual analysis since there is only one way to control the states to $(0, 0)$, so that way must be optimal among "all" the possibilities.

Essentially, the fixed time removes the infinitesimal approach to the $x_1$-axis in the usual energy-optimal problem. The points that can reach the origin in 5 or fewer seconds can thus all be optimally controlled to $(0, 0)$ while minimizing energy.

**Lab Activity 9 (solutions)**

We use the $u^* = 1$ trajectory from $(-1, -1)$,

$$x_1(t) = \frac{-2 - 2t + t^2}{2} \quad \text{and} \quad x_2(t) = t - 1,$$

until intersecting the $u^* = 0$ part of the switching curve $x_1 = -\frac{5}{2}x_2^2$ when $t = 1 + \frac{1}{\sqrt{2}}$ at $(x_1, x_2) = (-\frac{5}{4}, \frac{1}{\sqrt{2}})$. We then use this point as the initial state for the $u^* = 0$ trajectory

$$x_1(t) = \frac{1}{4}(-5 + 2\sqrt{2}t) \quad \text{and} \quad x_2(t) = \frac{1}{\sqrt{2}},$$

which intersects with the $u^* = -1$ part of the switching curve

$$sc_1(s) = -\frac{s^2}{2} \quad \text{and} \quad sc_2(s) = -s$$

when $t = \sqrt{2}$ and $s = -\frac{1}{\sqrt{2}}$. The total time from $(-1, -1)$ to $(0, 0)$ is thus

$$1 + \frac{1}{\sqrt{2}} + \sqrt{2} + \left| -\frac{1}{\sqrt{2}} \right| = 3.828.$$

Here is an image of the trajectories in the $x_1 x_2$-plane:

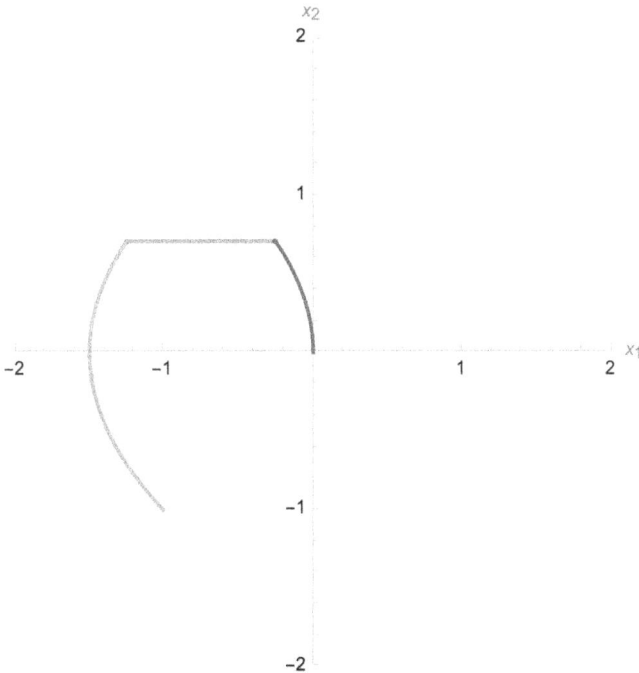

# Bibliography

[1]  R. Bellman. Eye of the Hurricane. World Scientific Publishing Co Pte Ltd., Singapore, 1984.

[2]  E. R. Pinch. Optimal Control and the Calculus of Variations. Oxford University Press Inc., New York, 1993.

[3]  M. Mesterton-Gibbons. A Primer on the Calculus of Variations and Optimal Control Theory. The American Mathematical Society, Providence, Rhode Island, 1990.

[4]  V. M. Tikhomirov. Stories About Maxima and Minima. The American Mathematical Society, Providence, Rhode Island, 1990.

[5]  W. Schaffer. The application of optimal control theory to the general life history problem. *The American Naturalist* **121**:418–431, 1983.

[6]  W. Schaffer, R. Inouye, and T. Whittam. The dynamics of optimal energy allocation for an annual plant in a seasonal environment. *The American Naturalist* **120**:787–815, 1982.

https://doi.org/10.1515/9783111290157-022

# Index

https://doi.org/10.1515/9783111290157-023

www.ingramcontent.com/pod-product-compliance
Lightning Source LLC
Chambersburg PA
CBHW081532220326
41598CB00036B/6413

* 9 7 8 3 1 1 1 2 8 9 8 3 0 *